Hauke Gergeleit

Zytologische und bakteriologische Untersuchungen von Sekreten der Sinūs paranasales des Pferdes

Cuvillier Verlag Göttingen
Internationaler wissenschaftlicher Fachverlag

Bibliografische Information der Deutschen Nationalbibliothek
Die Deutsche Nationalbibliothek verzeichnet diese Publikation in der Deutschen Nationalbibliografie; detaillierte bibliographische Daten sind im Internet über http://dnb.d-nb.de abrufbar.

1. Aufl. - Göttingen: Cuvillier, 2019
Zugl.: Hannover (TiHo), Univ., Diss., 2019

Nonnenstieg 8, 37075 Göttingen
Telefon: 0551-54724-0
Telefax: 0551-54724-21
www.cuvillier.de

1. Auflage, 2019
Gedruckt auf umweltfreundlichem, säurefreiem Papier aus nachhaltiger Forstwirtschaft.

ISBN 978-3-7369-7021-2
eISBN 978-3-7369-6021-3

Tierärztliche Hochschule Hannover

Zytologische und bakteriologische Untersuchungen von Sekreten der Sinūs paranasales des Pferdes

INAUGURAL – DISSERTATION
Zur Erlangung des Grades eines Doktors der Veterinärmedizin
- Doctor medicinae veterinariae -
(Dr. med. vet.)

vorgelegt von

Hauke Gergeleit

Pinneberg

Hannover 2019

Wissenschaftliche Betreuung:

Prof. Dr. Bernhard Ohnesorge
Klinik für Pferde
Tierärztliche Hochschule Hannover

PD Dr. habil. Astrid Bienert-Zeit
Klinik für Pferde
Tierärztliche Hochschule Hannover

1. Gutachter: Prof. Dr. Bernhard Ohnesorge

2. Gutachter: Prof. Dr. Martina Hoedemaker, PhD

Tag der mündlichen Prüfung: 16. Mai 2019

Meiner Mutter in großer Dankbarkeit

Meinem Vater in liebevollem Gedenken

Ergebnisse dieser Dissertation wurden in international anerkannten Fachzeitschriften zur Veröffentlichung angenommen:

- Journal of Equine Veterinary Science (akzeptiert am 02. November 2017):
  Cytologic and microbiological examination of secretions from the paranasal sinuses in horses and other species
  H. Gergeleit, A. Bienert-Zeit, B. Ohnesorge
  Volume 61, January 2018, Pages 22–31

- Acta Veterinaria Scandinavica (akzeptiert am 23. Juni 2018):
  A prospective study on the microbiological examination of secretions from the paranasal sinuses in horses in health and disease
  H. Gergeleit, J. Verspohl, J. Rohde, C. Rohn, B. Ohnesorge, A. Bienert-Zeit
  Volume 60, Issue 1, 2018, Pages 43–51

- Journal of Equine Veterinary Science (akzeptiert am 10. März 2019):
  Cytological examination of secretions from the paranasal sinuses in horses
  H. Gergeleit, A. Bienert-Zeit, A. Seemann-Jensen, J. Delarocque, B. Ohnesorge
  Volume 78, July 2019, Pages 60–68

- Pferdespiegel
  Sekrete aus den Nasennebenhöhlen des Pferdes – Untersuchungsmöglichkeiten und deren diagnostische Aussagekraft
  H. Gergeleit, B. Ohnesorge, A. Bienert-Zeit
  Pferdespiegel 2019; 22: Seiten 3–9

Teilergebnisse dieser Dissertation wurden auf folgenden Fachkongressen präsentiert:

- Equine Veterinary Dental Forum, 31.05.-02.06.2018, Innsbruck/ Österreich
  Microbiological examination of secretions from the paranasal sinuses in horses
  H. Gergeleit , A. Bienert-Zeit, J. Verspohl, J. Rohde, K. Rohn, B. Ohnesorge

- Zahntag des bpt Kongresses, 15.11.2018, Hannover
  Bakteriologische und zytologische Befunde aus Sekreten der Nasennebenhöhlen des Pferdes
  H. Gergeleit, A. Bienert-Zeit, B. Ohnesorge

Inhaltsverzeichnis

1 Einleitung 1

2 Literaturübersicht 3

2.1 Publikation I 3

3 Ergebnisse 15

3.1 Publikation II 15

3.2 Publikation III 26

4 Übergreifende Diskussion 37

4.1 Diskussion von Material und Methode 37

4.1.1 Probandengut 37

4.1.2 Probengewinnung 38

4.1.3 Probenauswertung 39

4.2 Diskussion der Ergebnisse 40

4.2.1 Ergebnisse der mikrobiologischen Untersuchung 40

4.2.2 Ergebnisse der zytologischen Untersuchung 44

4.2.3 Diagnostischer Nutzen der Sekretuntersuchung 45

5 Zusammenfassung 47

6 Summary 49

7 Literaturverzeichnis 51

8 Danksagung 55

9 Erklärungen über die erbrachten Eigenleistungen 57

Abkürzungsverzeichnis

BAL	bronchoalveoläre Lavage
BALF	bronchoalveoläre Lavage-Flüssigkeit
CT	Computertomographie, computertomographisch
IAD	inflammatory airway disease
NNH	Nasennebenhöhle(n)
PSH	Progressives Siebbeinhämatom
RAO	recurrent airway obstruction
rDNA	ribosomale Desoxyribonukleinsäure
S. zooepidemicus	*Streptococcus equi subspecies zooepidemicus*
spp.	species (Plural)
ssp.	subspecies
TBS	Tracheobronchialsekret

1 Einleitung

Entzündungen der Nasennebenhöhlen sind bei Pferden häufig beschriebene Erkrankungen. Sie werden meist erst im subakuten bis chronischen Krankheitsstadium diagnostiziert, da das Leitsymptom des einseitigen Nasenausflusses aufgrund der komplexen anatomischen Struktur der equinen Nasennebenhöhlen in der Regel erst bei fortschreitender Erkrankungsdauer auftritt (WAGUESPACK u. TAINTOR 2011; DIXON et al. 2012). Eine gesicherte Diagnose über die zugrundeliegende Ätiologie der Sinusitis erfordert eine aufwendige Kombination mehrerer bildgebender Verfahren (BEARD u. HARDY 2001).

Die wichtigsten diagnostischen Verfahren stellen derzeit, neben der klinischen Allgemeinuntersuchung und der speziellen Untersuchung des Kopfes, die transnasale Endoskopie der Nasenhöhle, der Nasennebenhöhlen sowie röntgenologische Untersuchungen des Kopfes dar. Die invasive Sinuskopie des Sinus conchofrontalis oder Sinus maxillaris rostralis beziehungsweise caudalis bietet nicht nur die Möglichkeit des direkten Einblicks in verschiedene Sinuskompartimente. Über die Trepanationsöffnungen können gegebenenfalls auch direkt Sekretproben oder Biopsien entnommen werden sowie therapeutische Maßnahmen erfolgen (RUGGLES et al. 1993). Darüber hinaus kommt der computertomographischen (CT) Untersuchung der Nasennebenhöhlen eine wesentliche diagnostische Bedeutung zu (HENNINGER et al. 2003), die nicht zuletzt die Wahl des chirurgischen Zugangs und der weiteren Therapie entscheidend erleichtern kann. Weitere am Pferdeschädel eingesetzte bildgebende Verfahren stellen die Szintigraphie und die Magnetresonanztomographie dar, die sich gegenüber der CT Untersuchung allerdings als weniger aussagekräftig herausgestellt haben (BARAKZAI et al. 2006; KAMINSKY et al. 2016).

Zytologische und bakteriologische Untersuchungen der Sekrete aus den Nasennebenhöhlen wurden zuvor nur sporadisch beschrieben und ihre ätiologische Relevanz blieb bislang ungeklärt (TREMAINE et al. 1999; COWELL u. TYLER 2001; TREMAINE u. FREEMAN 2007).

Dagegen liefert die zytologische Analyse des transendoskopisch gewonnenen Tracheobronchialsekrets bei Erkrankungen der tiefen Atemwege wichtige Aussagen über Art und Schwere der Erkrankung und dient darüber hinaus auch der prognostischen Einschätzung (DERKSEN et al. 1989; MAY u. GEHLEN 2009).

Vor diesem Hintergrund werden im Rahmen der eigenen Untersuchungen Sekretproben sowohl transendoskopisch als auch direkt aus den Sinus paranasales von gesunden und kranken Pferden entnommen, um diese zytologisch und bakteriologisch zu untersuchen. Ziel hierbei ist es zunächst herauszufinden, wie sich das physiologische Zellbild und die Keimflora des Pferdes in den Nasennebenhöhlen darstellen, um in einem zweiten Schritt festzustellen, ob sich aus den bei erkrankten Pferden erhobenen Befunden retrospektiv Hinweise auf die zugrundeliegende Erkrankung ableiten lassen. In der abschließenden Diskussion wird der Frage nachgegangen, ob und welchen diagnostischen Nutzen diese Untersuchungen bei der Evaluation von Patienten mit Erkrankungen der Nasennebenhöhlen haben können.

2 Literaturübersicht

2.1 Publikation I

Cytologic and microbiological examination of secretions from the paranasal sinuses in horses and other species

Published in: Journal of Equine Veterinary Science
Volume 61, January 2018, Pages 22–31

Received: 14 September 2017
Received in revised form: 01 November 2017
Accepted: 2 November 2017
Available online: 10 November 2017

H. Gergeleit, A. Bienert-Zeit, B. Ohnesorge
Clinic for Horses, University of Veterinary Medicine Hannover, Foundation
Bünteweg 9, 30559 Hannover, Germany

Abstract

Only sparse literature has been published on the microbiological and cytological examination of secretions from the paranasal sinuses in horses, and the data that is available consists mainly of results in single animals or small case series. Contrarily, the role of bacteria in sinusitis in humans has been the topic of several studies and reviews.

We aim with this review to summarize the current knowledge on the characteristics and microbiological and cytologic composition of paranasal sinus secretions in equids in comparison to humans and selected domestic animals.

Journal of Equine Veterinary Science 61 (2018) 22–31

Contents lists available at ScienceDirect

Journal of Equine Veterinary Science

journal homepage: www.j-evs.com

Review Article

Cytologic and Microbiological Examination of Secretions From the Paranasal Sinuses in Horses and Other Species

Hauke Gergeleit*[,1], Astrid Bienert-Zeit [1], Bernhard Ohnesorge

Clinic for Horses, University of Veterinary Medicine Hannover, Foundation, Hannover, Germany

ARTICLE INFO

Article history:
Received 14 September 2017
Received in revised form 1 November 2017
Accepted 2 November 2017
Available online 10 November 2017

Keywords:
Sinusitis
Horse
Cytology
Microbiology
Anaerobe

ABSTRACT

Only sparse literature has been published on the microbiological and cytologic examination of secretions from the paranasal sinuses in horses, and the data that are available consist mainly of results in single animals or small case series. Contrarily, the role of bacteria in sinusitis in humans has been the topic of several studies and reviews. We aim with this review to summarize the current knowledge on the characteristics and microbiological and cytologic composition of paranasal sinus secretions in equids in comparison to humans and selected domestic animals.

1. Introduction

Sinusitis is a common disorder in horses with different underlying etiologies [1,2]. Horses are mostly presented in a chronic stage of the disease [3], and an exact etiologic diagnosis may only be achieved by a combination of different diagnostic imaging tools.

There have been quite a lot of retrospective studies, case series, and reviews in recent years regarding historical and clinical features, ancillary diagnostic techniques [1–6], medical and surgical therapy, and long-term results of treatment in horses with sinusitis [7,8]. However, the role of the cytology and microbiology of secretions associated with sinusitis in horses has only been sporadically reported [9–13] or was of unknown importance [11,14].

The purpose of this report is to review pathophysiologic conditions in sinusitis and to summarize findings regarding the cytology and microbiology of secretions to evaluate their diagnostic value.

2. Anatomic and Pathophysiologic Considerations in Sinusitis

2.1. Pathophysiology of Sinusitis

The paranasal sinuses of the horse are air-filled cavities that extent bilaterally within both sides of the skull. Theories on their physiological function include lightening of the skull, humidification, and warming of the inspired air as well as moistening of the nasal olfactory mucosa. All these theories are deniable to some extent so that their precise role remains uncertain [2,15]. Sinusitis is a state of inflammation of the paranasal sinus mucosa [16]. The basic pathophysiological mechanism in the development of sinusitis is the ineffectiveness of mucociliary clearance. The role of mucociliary clearance is to remove physiological and pathologic secretions from the airway and is, therefore, considered to be the primary defense mechanism of the paranasal sinuses [17]. The paranasal sinuses are covered

Animal welfare/ethical statement: This work does not contain own research results on animals.

Conflict of interest statement: The authors declare no conflicts of interest.

* Corresponding author at: Hauke Gergeleit, Clinic for Horses, University of Veterinary Medicine Hannover, Foundation, Bünteweg 9, 30559 Hannover, Germany.

E-mail address: hauke.gergeleit@tiho-hannover.de (H. Gergeleit).

[1] These authors are joint first authors.

0737-0806/$ – see front matter © 2017 Elsevier Inc. All rights reserved.
https://doi.org/10.1016/j.jevs.2017.11.001

by respiratory pseudostratified, columnar, ciliated epithelium, and the latter contains fewer goblet cells and submucosal seromucous glands than the epithelium in the nasal cavity [14]. These glands generate the mucus layer that covers the epithelium. The cilia create a mucus movement to the natural ostia [18].

The beating cilia and specific physical characteristics of the overlying mucus determine the clearance of the physiological and pathologic substances in the mucus. The continuous and coordinated beat of the respiratory cilia is modulated dynamically and can increase to accelerate mucus clearance in times of stress, such as exercise, infection, or fever [17].

The restoration of natural sinus physiology became the goal of medical and surgical treatment of sinonasal inflammatory disease in humans from the time Messerklinger described the sinonasal mucociliary flow patterns in 1966 [17]. He determined that the transport of secretions from the paranasal sinuses is dependent on five factors: (1) The excretory pathways; (2) the secretion; (3) the features of the secretion; (4) the cilia beat; and (5) the resorption [19].

If inflammation occurs, dilation of the venous sinusoids in horses can lead to edematous swelling of the sinus mucosa of up to 15 mm [14] (Fig. 1). Consequent narrowing of the ostia between the sinuses and especially the nasomaxillary aperture can obstruct the drainage pathways completely. This contributes to an intrasinusoidal accumulation of secretions, therefore, impairing local mucosal defense mechanisms further and promoting progression of underlying disease [14,17,18,20]. These conditions predispose for a bacterial infection [14]. The initial seromucous secretion becomes more viscous with the increase in inflammatory cell extravasation. Inflammatory products that cannot be eliminated tend to accumulate, dry out, and promote a chronic inflammatory state [16].

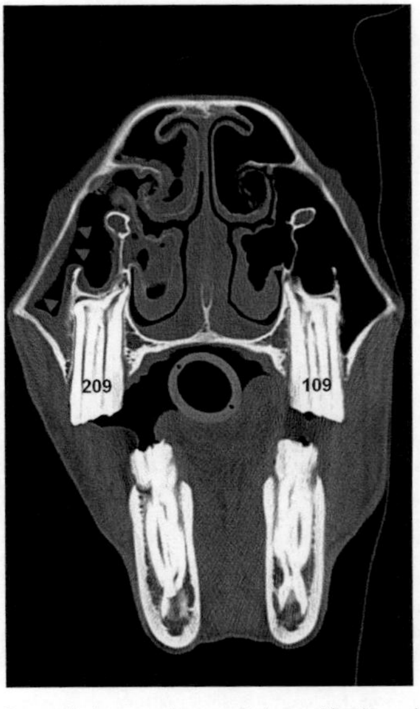

Fig. 1. Computerized tomography scan of a horse with primary sinusitis. Note, the swollen mucosa of the diseased maxillary sinus (red arrowheads, left) compared to the nonaffected side (right). The horse is positioned in right lateral recumbency.

2.2. Sinusitis in Horses

The equine paranasal sinuses represent an anatomically complex, yet well-described system [3,21–27]. The dependency for drainage of secretions of the adjacent paranasal sinuses on the rostral and caudal maxillary sinus [4,25], in addition to the physiologically very narrow nasomaxillary aperture [22], predisposes horses to develop paranasal sinusitis with inspissation of secretions [4,28–30]. In a prolonged clinical course of disease, the inspissated pus can have a caseous character that is only rarely resolved without surgical intervention [28–30].

The most common clinical feature of paranasal sinusitis in horses is unilateral nasal discharge [1,5,31,32] (Fig. 2), followed by ipsilateral mandibular lymphadenitis. However, complete obstruction of the drainage pathways can result in absence of characteristic nasal discharge [6]. Etiologically, paranasal sinus diseases can be classified as primary, with a viral, bacterial, or mycotic origin, or secondary to dental diseases, sinus cysts, neoplasia, and trauma [1,2].

The character of the nasal discharge in horses with primary and dental sinusitis is usually mucopurulent or purulent [1]. In primary sinusitis, it is usually odorless [9,32], but can sometimes be malodourous [6,32]. Schumacher et al [28] found foul-smelling unilateral nasal discharge in five horses with primary sinusitis that was associated with inspissated purulent exudate in the ventral conchal sinus. Most horses with dental sinusitis show malodorous nasal discharge [4,5,12,32,33]. The fetid smell usually suggests the involvement of obligate anaerobic bacteria, which is ascribable to the production of butyric acid in the metabolism of anaerobes [34].

A translucent, honey-colored, odorless, viscous nasal discharge can indicate a sinus cyst (Fig. 2B), which can become more purulent with the occurrence of secondary infection [5,6].

Hemorrhagic nasal discharge or epistaxis is more common with traumatic sinusitis and is typical for progressive ethmoidal hematoma (PEH) [5,6]. The discharge is usually odorless.

It is important to note that unilateral nasal discharge does not necessarily originate from the paranasal sinuses. The nares, the nasal cavity, the guttural pouches, and the lungs display anatomic locations which have to be taken into consideration when a suspected horse is presented [16,35–37].

2.3. Sinusitis in Humans

Most infections of the paranasal sinuses in human medicine are rhinogenic, spreading from the nose into the

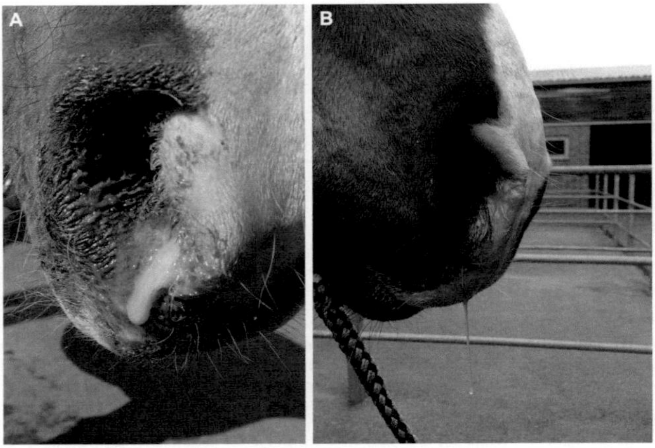

Fig. 2. Unilateral nasal discharge: purulent in a horse with dental sinusitis (A) and translucent, honey colored in a horse with a sinus cyst (B).

sinuses [20]. In contrast to the horse, a sinusitis in humans rarely occurs alone but is mostly associated with inflammation of the adjacent nasal mucosa [38]. Therefore, the term rhinosinusitis replaced the term sinusitis in many expert panels [38–41]. Acute rhinosinusitis (ARS) is usually characterized as a viral upper respiratory tract infection with typical symptoms, such as nasal blockage, discharge, facial pressure, and reduction in the ability to smell [42] lasting 4 weeks or less [38,40,43]. The term chronic rhinosinusitis (CRS) is used by most guidelines if symptoms last 12 weeks or longer [38–40,42]. The estimated incidence of ARS is two to five episodes per year in an average adult [42]. Rhinosinusitis is highly prevalent and constitutes an enormous economic burden [44,45].

Only 0.5%–2% of ARS cases are thought to be complicated by a secondary bacterial infection [39], making proper identification of cases with bacterial etiology one of the primary challenges in the treatment of ARS [44]. It is estimated in the pediatric population that secondary bacterial infection occurs in 5%–13% of viral upper respiratory tract infections [46]. Neither nasal mucus color nor the presence of fever is useful to differentiate bacterial from viral etiology [38,43].

About 10% of sinusitis cases have an odontogenic origin [47], including endodontic infections, periapical pathologies, and iatrogenic complications [48]. In fact, the majority of these cases (up to 64%) develop secondary to dentoalveolar surgical interventions with orosinuidal fistula formation [49], often associated with dislocated foreign bodies in the maxillary sinus [50].

2.4. *Sinusitis in Other Domestic Animals*

Comparable to humans, rhinosinusitis in cats and dogs occurs more frequently than an isolated inflammation of the parasinusoidal mucosa. The most common cause of chronic mucoid to hemorrhagic nasal discharge in dogs is sinonasal aspergillosis with Aspergillus fumigatus being the species most frequently encountered [51]. The most common causes of chronic nasal disease in cats are idiopathic CRS and neoplasia [52].

The same etiologies for paranasal sinusitis are described in cattle as those in horses [53–56], although they are reported relatively infrequently. Frontal sinusitis often occurs secondary to dehorning [55].

3. Access to the Paranasal Sinuses and Specimen Collection

An upper airway endoscopy is the second diagnostic step after the physical examination if sinusitis is suspected [5]. It is easy to perform in restrained or sedated horses, displays a noninvasive procedure, and is of high diagnostic value in most cases. The anatomic structures of the nasal cavity and the nasomaxillary aperture are usually easily identified. The typical endoscopic appearance of horses affected by sinusitis is discharge from the caudal or rostral maxillary sinus down the drainage angle between the ventral and dorsal nasal concha at the end of the middle nasal meatus via the nasomaxillary aperture (Fig. 3). Transnasal access to the paranasal sinuses is also sometimes possible using a flexible endoscope with a diameter <5 mm, but this requires more experience and is seldom possible for the rostral paranasal sinuses [25,57].

Specific endoscopic diagnosis is aggravated if severe narrowing of the nasal passages restricts proper advancement and positioning of the endoscope due to expansion of the nasal conchal sinuses [5]. The nasomaxillary aperture can be blocked completely by mucosal inflammation,

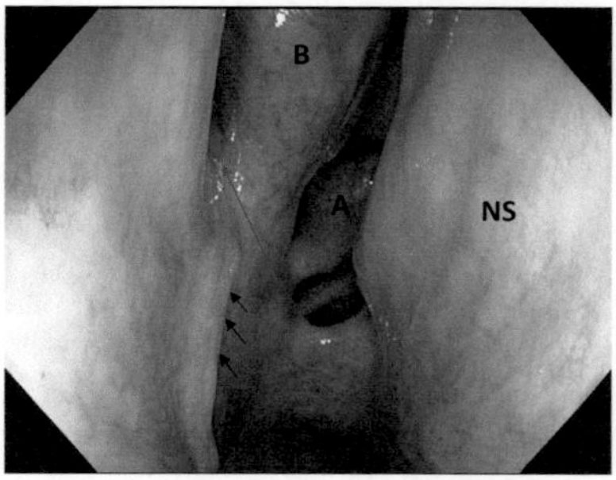

Fig. 3. Endoscopic image the "drainage angle" (red arrow) with purulent, malodorous discharge (black arrows) in a horse with squamous cell carcinoma of the right sinus system. A—middle nasal concha; B—dorsal nasal concha; NS— nasal septum.

resulting in the retention of secretions and absence of characteristic discharge.

Trephination of the paranasal sinuses allows direct access to gain samples of secretions for further evaluation but is obviously a more invasive procedure. It is usually performed with a therapeutic intention but may also have a diagnostic value when a sinuscopy is performed. Although the surgical and cosmetic outcome is most commonly satisfying, opening of the paranasal sinuses holds the risk of wound infection and delayed secondary healing [58].

Anterior rhinoscopy is the first step in examining humans affected by rhinosinusitis, albeit of limited value. Better visualization is achieved by nasal endoscopy [42]. Maxillary sinus puncture has been the traditional gold standard in human medicine to assess the microbial flora associated with bacterial rhinosinusitis [59]. Pursuing alternative harvesting methods with less patient discomfort and minimized risks, nasal cultures, and endoscopically guided middle meatal (EDMM) cultures have been compared to maxillary sinus taps. However, nasal cultures have poor correlation to maxillary sinus cultures [60]. By contrast, studies have shown that there is a satisfying concordance between EDMM cultures and maxillary sinus cultures [60–63], but their validity is limited by small sample sizes. Nevertheless, Benninger et al [64] concluded EDMM cultures to be a feasible, highly sensitive, and accurate culture method for bacterial rhinosinusitis in humans.

4. Microbiology of Secretions From the Paranasal Sinuses

4.1. Microbiology of Secretions From the Paranasal Sinuses in Horses

Despite the paucity of studies focusing on the role of bacteria in equine paranasal sinus diseases, there are some studies and case reports where cultures of exudates from sinusitis have been examined for bacterial growth (Table 1). A wide variety of potential bacterial pathogens was reported to be found on sinusitis mucosae and their etiologic importance often remained unclear [6,14,33].

Anaerobes seem to play an important role in sinusitis with dental origin. They have been found to be part of the normal flora of the respiratory tract and the oral cavity [10,65,66]. The close relationship of the caudal cheek teeth to the maxillary sinuses makes the latter prone to become secondarily infected by dental diseases [66]. These diseases often reveal mixed bacterial growth, with Gram-negative obligatory anaerobes being the predominant group differentiated, with *Prevotella* spp., *Fusobacterium* ssp., and *Bacteroides* ssp. being isolated most frequently [12,67,68]. *Prevotella* and *Veillonella* were the dominating genera in a study on the microbiome associated with equine periodontitis [66]. *Streptococcus* spp. are the aerobic organisms most commonly associated with periodontal diseases [67,68].

Only one publication reports the presence of anaerobes in a case of primary sinusitis [28]. Most of them report mixed aerobic bacteria or pure cultures of *Streptococcus* spp. [5,6,13].

Equine sinus cysts are described as being most commonly bacteriologically sterile [5,14], but secondary bacterial infection may occur [6]. Cysts in horses show similarities to human mucoceles [69].

Table 1
Microbiological findings in horses diagnosed with sinusitis.

Diagnosis	Aerobic and Facultative Bacteria		Anaerobic Bacteria
	Gram Positive	Gram Negative	
Dental sinusitis	*Staphylococcus aureus* [9] β-hemolytic *Streptococci* [9] *Actinomyces* spp. [12]	*Escherichia coli* [9] *Diplococcus mucosus* [9]	*Bacteroides* spp. [10] *Fusobacterium* spp. [10,12] *Peptostreptococcus* spp. [10] *Prevotella* spp. [12]
Primary sinusitis	*Streptococcus* equi var. equi [6,13] Strep. equi var. zooepidemicus [6,27] *Staphylococcus* spp. [27] *Corynebacterium* spp. [28]	Mixed growth [27]	*Bacteroides melaninogenicus* [27]
Sinusitis (nonspecific)	*S. aureus* [9] β-hemolytic *Streptococci* [9,11]	*E. coli* [11] *Pseudomonas* spp. [9,11]	*Bacteroides* sp. [11] *Peptostreptococcus* spp. [11]

An etiologic diagnosis could not be made for all cases.
Tremaine and Dixon (2001) [6]; Mason (1975) [9]; Mackintosh and Colles (1987) [10]; Ruggles et al (1993) [11]; Bartmann et al (2002) [12]; Laverty and Pascoe (1997) [13]; Schumacher et al (1987) [28]; Schumacher and Crossland (1994) [29].

Mycotic pathogens may sometimes be responsible for sinusitis in horses. Aspergillus fumigatus has been described as a common cause of mycotic rhinitis in Great Britain and was also found as a frequent cause of sinonasal mycosis, but heavy growth of other potentially pathogenic fungi is also relevant [6]. Aetiopathogenetically, *Cryptococcus neoformans, Coccidiodes immitis, Allescheria boydii*, and *Pseudallescheria boydii* have been reported to sporadically cause granulomas in the equine nasal passages and paranasal sinuses [70–74], but these have a more distinct geographical distribution in warmer climates, such as Australia and North America [14]. Transient mycotic infections can occur after sinus surgery [6,14] and after long-term antibiotic administration.

4.2. Microbiology of Secretions From the Paranasal Sinuses in Humans

Bacteria isolated from both healthy and diseased paranasal sinuses have been widely studied in human medicine. Although the most common pathogens for acute and chronic bacterial sinusitis seem to have been identified, there is still some debate about the physiological flora of the paranasal sinuses. Brook [75] reported primarily anaerobic bacteria from normal maxillary sinuses and Jiang et al [76] concluded endoscopically normal maxillary sinuses not to be sterile, whereas other authors described the maxillary or frontal sinuses to be mainly sterile, respectively [77–79]. Kremer et al reported coagulase-negative staphylococci in almost all their samples taken from healthy individuals and found *Staphylococcus aureus* being present in up to one-third of them. However, based on their bacteriological findings, no differentiation could be made between the control group and patients with chronic sinusitis [80].

The predominant organisms found in acute sinusitis are Haemophilus influenzae, *Streptococcus pneumoniae*, and *Moraxella catarrhalis* [64,81–84]. Pure cultures of one pathogen were found in up to 56% of positive samples [82]. Bacteria that were identified less frequently include *S. aureus* and other *Staphylococci*, *Streptococci*, and *Pseudomonae* as well as Enterobacteriaceae [81–84]. *Staphylococcus aureus* and some other species are rather seen as contaminants from the nasal cavity [82]. Obligatory anaerobe bacteria, including *Bacteroides* spp., *Peptostreptococcus* ssp., and *Fusobacterium* ssp., were infrequently isolated in samples, with their occurrence ranging between 0% and 30% [81,83,84].

A greater variation of bacteriological cultures exists in chronic sinusitis [60,81]. Although the organisms found are similar to those in acute sinusitis, the prevalence and quantity when detecting them are significantly higher as well as the occurrence of polymicrobial cultures [81].

Table 2 summarizes some of the bacteria identified more frequently found in humans with chronic sinusitis. These studies differ in their sample collection methods and display conflicting results, making it very difficult to compare the results of various authors. Furthermore, the quantity of bacterial growth was not consistently described.

Mantovani et al [85] collected aspirates from the maxillary sinus punctures in patients with CRS. No bacterial

Table 2
Selection of microbiological findings described frequently in humans diagnosed with chronic sinusitis [68,77–82].

Aerobic and Facultative Bacteria		Anaerobic Bacteria
Gram Positive	Gram Negative	
Staphylococcus aureus	*Pseudomonas aeruginosa*	*Bacteroides* ssp.
Coagulase-negative *Staphylococcus* ssp.	*Enterobacteriaceae*	*Fusobacterium* ssp.
Streptococcus spp.		*Peptostreptococcus* ssp. *Veillonella* spp. *Propionibacterium* spp.

growth was found in 53.2% of the samples, and no anaerobes were detected at all, whereas 45.2% revealed aerobic bacteria and one sample was positive for *C. neoformans*. *Pseudomonas aeruginosa* was identified as the most frequently occurring (27.6% of positive cultures), followed by other Gram-negative bacteria (31%) and *Staphylococcus* spp. (27.6%), and the authors concluded that they represented the bacterial flora found in CRS patients. These results are similar to those from other studies, which also found *P. aeruginosa* and *S. aureus* being cultured most frequently in patients with CRS [86]. *Pseudomonas aeruginosa* was identified more commonly in patients who had undergone functional endoscopic sinus surgery [87].

Rombaux et al [88] also found *Staphylococcus* ssp. and Gram-negative aerobes in more than 50% of their cases. Enterobacteriaceae had a high incidence in positive cultures (43.7%), underlining their importance in chronic sinusitis. Mantovani et al [85] regarded Enterobacteriaceae as pathogenic agents with a secondary role.

Microbiological detection of anaerobes varies widely between studies. Doyle and Woodham [89] found no anaerobes when they cultured biopsies from inflamed ethmoid sinuses, whereas Brook [90] detected anaerobes in 88% of the samples from chronically inflamed maxillary sinuses, which were dominated by anaerobic cocci and *Bacteroides* spp. Other studies found anaerobes in 2% and 14% of positive cultures, respectively, including *Bacteroides* spp., *Peptostreptococcus* spp., *Veillonella* spp., *Fusobacterium* spp., and *Propionibacterium* spp. [88,91]. Maxillary sinusitis associated with an odontogenic origin is generally polymicrobial [92], and it is predominated by anaerobic bacteria, including Gram-negative bacilli, *Peptostreptococcus* spp., and *Fusobacterium* spp. in both acute and chronic stages [93].

4.3. Microbiology of Secretions From the Paranasal Sinuses in Other Domestic Animals

Nasal swabs from healthy dogs were dominated by *Moraxella* spp., *Phyllobacterium* spp., *Cardiobacterium* spp., and *Staphylococcus* spp. In diseased dogs, *Moraxella* spp. were significantly less frequently identified, whereas Pasteurellaceae were found significantly more often [94].

Pseudomonas spp. were most commonly found in samples collected from the frontal sinus during surgery in cats with chronic nasal discharge [95] and from nasal brushings in cats with nonspecific chronic nasal disease [96]. Other

H. Gergeleit et al. / Journal of Equine Veterinary Science 61 (2018) 22–31 27

bacteria cultured included *Escherichia coli*, *Staphylococcus* spp., *Streptococcus* spp., *Pasteurella* spp., *Serratia* spp., *Klebsiella* spp., and *Proteus* spp [96]. Potentially pathogenic bacteria, anaerobic bacteria, and *Mycoplasma* spp. are isolated more often in cats with idiopathic CRS, and the variety of bacterial species found is higher than in healthy cats [97]. Another study revealed *Moraxella* spp. being cultivated most frequently in healthy cats and individuals affected by cat flu. Bradyrhizobiaceae dominated the bacterial microbiome of cats with nasal neoplasia [98].

Bovine paranasal sinuses of clinically healthy individuals were predominantly sterile or did not contain microbes that are detectable with standard cultivating methods. Cultivation of swab samples taken from different sinuses after trephination sporadically revealed *Mannheimia haemolytica*, *Trueperella pyogenes*, *Pasteurella multocida*, and *Bibersteinia trehalosi*, among others [99]. *T. pyogenes* and *P. multocida* were detected in cases of chronic frontal sinusitis [55].

4.4. Assessment of the Microbiome

Studies have shown that many bacteria cannot be cultured using conventional methods due to nutritional and fastidious growth requirements [100]. It has become possible to detect almost the entire community of commensal and pathogenic bacteria of individuals and populations culture independently by using broad platform molecular techniques. This universe of cultivable and uncultivable microorganisms present in a specific ecologic niche, such as the sinus mucosa, is called the microbiome. The assessment of the microbiome in human medicine is already highly prevalent, and sequencing of the bacterial 16S ribosomal RNA has become the standard for identification and phylogenetic classification [101]. The advantages include far greater coverage, accuracy, and sensitivity compared to conventional cultivating methods, allowing a better understanding of host-microbial relations. Wilson and Hamilos [102] summarized the current knowledge on the nasal and sinus microbiome in human CRS and healthy controls. In their review, they highlight the increased burden of certain organisms (e.g., *S. aureus*) associated with CRS. The total bacterial burden in CRS patients is described as similar to healthy individuals. They also refer to studies that revealed anaerobic bacteria to be the most prevalent organisms in CRS patients using pyrosequencing [100,103] and express doubts regarding the validity of earlier culture-based studies that reported a low prevalence of anaerobes in CRS samples.

5. Histological Features of Biopsies From the Paranasal Sinuses in Horses

There are no studies that focus on the cytologic examination of secretions that originate directly from the equine paranasal sinuses.

A large number of neutrophils were found in the nasal secretions from 30 horses with infectious respiratory diseases, especially in those that were in the chronic stage of the disease, whereas eosinophils are the dominant cell population in allergic diseases [104].

Tremaine et al [14] examined biopsies from 79 horses with sinusitis histologically and found mostly chronic inflammatory changes, including mucosal thickening, loss of epithelial integrity, inflammatory cell infiltration, and significant fibroplasia in the sinus mucosa. The mucosal histological features of dental sinusitis were similar to those observed in equine chronic primary sinusitis, including mild-to-moderate infiltration of macrophages and predominantly moderate-to-severe infiltration of neutrophils. Overall, there was no significant correlation between the density of the cellular infiltrate and the severity of the disease [14,105].

Sinus cysts walls showed a mild, chronic, predominantly mononuclear, inflammatory infiltrate and, therefore, indicate a continuous low-grade inflammation [106,107]. Cyst aspirates contained 52%–99% neutrophils [14].

Histological examinations of PEHs often revealed ulcerated stratified squamous epithelium that showed a varying degree of fibroplasia with a localized underlying inflammatory infiltrate of plasma cells, lymphocytes and, less commonly, neutrophils. Numerous siderophages, a few eosinophils and numerous multinucleate giant cells, were sometimes present [14].

6. Cytology of Secretion From the Paranasal Sinuses in Humans and Domestic Animals

Studies in human medicine comparing the results of nasal cytology with those of sinus radiography had good sensitivity in predicting sinus disease using neutrophils counts [108,109]. However, these studies lack the comparison of nasal cytology to sinus aspiration [110].

Araujo et al [91] compared middle meatus secretions collected endoscopically in patients with CRS and healthy individuals and concluded that the microbiology of the middle meatus is similar in both groups. They pointed out the importance of the number of leucocytes to distinguish an infective from a saprophytic organism, as they found only rare or no leucocytes in the samples of the control group compared to numerous leucocytes in the patient group.

Bhattacharyya examined the cytology and microbiology of persistent paranasal sinus secretion in patients who had undergone functional endoscopic sinus surgery and in a control group. The control group demonstrated a higher amount of detached epithelial cells, whereas the diseased group had higher scores of neutrophils, but these results did not reach statistical significance. There was a statistically significant increase in eosinophilic cells in the absence of positive bacterial culture in the diseased group, whereas the number of neutrophils was found to be more distinct when bacteria were present [111].

Gupta et al [112] concluded fine-needle aspiration cytology to be a reliable diagnostic tool in numerous inflammatory and neoplastic sinus conditions, but it did not lead to a diagnosis in 12.6%.

Overall, it is concluded in the *European position paper on rhinosinusitis* that cytology has not been proven to be an adequate tool in the diagnosis of rhinosinusitis but that a biopsy might be indicated to exclude severe and malign conditions such as vasculitides and neoplasia [42].

Nasal mucosal cytology in cats and dogs has been shown to have poor sensitivity and specificity compared to histology in nasal mucosal samples and is often unreliable for detecting chronic changes but may be used to identify fungal or neoplastic lesions in selected cases [96,113].

7. Discussion

Paranasal sinusitis in horses is a well-described condition. So far, only a little or no diagnostic focus was put on the role of the microbiology and cytology of secretions of the paranasal sinuses. There are some case reports, but no comparative studies are published. The clinical significance often remained unclear in those cases reporting microbiological or cytologic findings. Consequently, the diagnostic value is either unknown, limited, or underestimated.

Several studies have been published on the role of microbiological findings in human medicine and some agree with findings in veterinary medicine. To date, there is no consensus on the physiological flora of the paranasal sinuses in humans, and it has not yet been assessed for equids. The spectrum of bacteria in equine primary sinusitis and human ARS include a comparably small number of pathogens differentiated, with some reports of pure cultures. *Streptococcus* species are frequently encountered in equids and humans. Although the main infective agents in ARS have been isolated, there is still no consensus on the bacterial flora in CRS as results vary widely between studies. Nevertheless, there is a tendency in humans and small animals toward a more diverse microbial profile in chronic diseases with higher incidences of *Enterobacteriaceae* and *Pseudomonas*. It seems likely that microbial examinations of horses with sinusitis will show similar results. One should also consider that horses affected by sinusitis are often presented to the referral veterinarian in a chronic stage of the disease and have usually already received antibiotics previously without permanent resolution of the symptoms [1]. These facts are likely to cause a shift in the microbiological composition and complicate the diagnosis and appropriate treatment.

Sinusitis with an odontogenic origin plays an important role in equine medicine but has a smaller incidence in humans, where it often occurs secondarily to dentoalveolar surgical interventions. However, there is little doubt that Gram-negative obligatory anaerobes dominate the bacterial spectrum in dental sinusitis in humans and have also been described in horses. Otherwise, their occurrence has only sporadically been described for equine sinusitis, other than when associated with a dental disease. The cultivation of anaerobes is more delicate, expensive, and labor intensive and might, therefore, be used less frequently in general practice and is possibly more prone to false-negative results due to fastidious growth requirements.

The range of detecting anaerobes in both healthy and diseased human individuals is enormous. Harvesting techniques vary between studies making direct comparisons, especially of microbial findings, difficult. Direct sinus punctures and protected swab insertion and sampling have only a small risk of contamination. However, it is a minimal invasive method and holds the general risks of complications. Comparative studies in human medicine conclude that endoscopically guided aspiration has a higher contamination rate by the endoscope itself due to insufficient cleaning or during nasal passage or unclean handling of the utensils, but the findings did not reach statistical significance [86]. Moreover, studies have shown good concordance between maxillary sinus puncture and EDMM cultures [60,64]. This approach could lead to the use of endoscopically guided transnasal culture methods in horses, avoiding opening of the paranasal sinuses before there is a surgical necessity. Nevertheless, it is essential to note that the assessment and appropriate treatment of the bacteria involved often does not solve the underlying disorder, as many cases of equine sinusitis are secondary to other diseases [6]. Therefore, identification of the bacteria may be used additionally to accurate clinical diagnostics and shall not delay the treatment of the underlying condition.

As previously described, assessment of the microbiome provides greater coverage, accuracy, and sensitivity than conventional cultivating methods [102]. Findings in human medicine indicate that alterations in the microbial community are more likely to be representative for a disease than infections with individual pathogens [100,103]. Notwithstanding this, culture-independent investigations have not yet been securely established in veterinary medicine. They have been used to determine the nasal microbiome in healthy and diseased cats and dogs [94,98] and to determine the microbiome associated with equine periodontitis and oral health [66]. To date, this approach has not been utilized to assess the nasal or paranasal microbiome in horses. Although the technologies are available, the methodical and financial effort is currently likely to exceed the clinical advantage.

It is concluded that the diagnostic value of cytologic examination of secretions from the paranasal sinuses and the nasal cavity is very limited in both human and small animal medicine. Nevertheless, some authors suggest that sample collection for cytologic examination in horses affected from sinusitis [5,114], but they do not comment on their interpretation. There are examples where analysis of secretions in airway diseases has been proven to be a useful tool. Cytologic analysis of tracheobronchial secretion and bronchoalveolar lavage fluid in horses, for instance, helps to confirm a suspected diagnosis by giving further information on character and severity of pulmonary diseases [115,116]. Bronchoalveolar lavage represents a sensitive method to diagnose diffuse diseases, such as chronic obstructive bronchitis and interstitial pneumopathies. Tracheobronchial secretion analysis is useful for generalized diseases by giving insight into the conditions of the respiratory epithelium and the effectiveness of mucociliary clearance [115]. These findings prompt one to suggest whether cytologic examination of paranasal sinus secretions might also be helpful to make an etiologic diagnosis and for follow-up examinations.

8. Prospects

Diagnostics in equine sinusitis can be challenging. In cases where an exact etiologic diagnosis remains unclear, even after clinical, endoscopical, and radiological examination, the evaluation of secretions collected transendoscopically could support a suspected diagnosis. The implementation of a computed tomography scan could possibly be avoided in such specific cases. However, studies focusing on the cytologic characteristics and bacteriological contents of secretions from the paranasal sinuses of healthy and diseased horses are necessary.

References

[1] Dixon PM, Parkin TD, Collins N, Hawkes C, Townsend NB, Fisher G, et al. Historical and clinical features of 200 cases of equine sinus disease. Vet Rec 2011;169:439.

[2] O'Leary JM, Dixon PM. A review of equine paranasal sinusitis. Aetiopathogenesis, clinical signs and ancillary diagnostic techniques. Equine Vet Educ 2011;23:148–59.

[3] Waguespack RW, Taintor J. Paranasal sinus disease in horses. Compend Contin Educ Vet 2011;33:E1–12.

[4] Dixon PM, Parkin TD, Collins N, Hawkes C, Townsend N, Tremaine WH, et al. Equine paranasal sinus disease: a long-term study of 200 cases (1997-2009): ancillary diagnostic findings and involvement of the various sinus compartments. Equine Vet J 2012;44:267–71.

[5] Beard W, Hardy J. Diagnosis of conditions of the paranasal sinuses in the horse. Equine Vet Educ 2001;13:265–73.

[6] Tremaine W, Dixon P. A long-term study of 277 cases of equine sinonasal disease. Part 1: details of horses, historical, clinical and ancillary diagnostic findings. Equine Vet J 2001;33:274–82.

[7] Dixon PM, Parkin TD, Collins N, Hawkes C, Townsend N, Tremaine WH, et al. Equine paranasal sinus disease: a long-term study of 200 cases (1997-2009): treatments and long-term results of treatments. Equine Vet J 2012;44:272–6.

[8] Tremaine W, Dixon P. A long-term study of 277 cases of equine sinonasal disease. Part 2: treatments and results of treatments. Equine Vet J 2001;33:283–9.

[9] Mason B. Empyema of the equine paranasal sinuses. J Am Vet Med Assoc 1975;167:727–31.

[10] Mackintosh M, Colles C. Anaerobic bacteria associated with dental abscesses in the horse and donkey. Equine Vet J 1987; 19:360–2.

[11] Ruggles AJ, Ross MW, Freeman DE. Endoscopic examination and treatment of paranasal sinus disease in 16 horses. Vet Surg 1993; 22:508–14.

[12] Bartmann C, Peters M, Amtsberg G, Deegen E. Dentogene Sinusitis durch gramnegative Anaerobier beim Pferd. Tierärztl Prax Großtiere 2002;30:178–83.

[13] Laverty S, Pascoe J. Sinusitis. Curr Ther Eq Med 1997;4:419–21.

[14] Tremaine W, Clarke C, Dixon P. Histopathological findings in equine sinonasal disorders. Equine Vet J 1999;31:296–303.

[15] Maroldi R, Nicolai P. Imaging in treatment planning for sinonasal diseases. Springer; 2005.

[16] Fey K, Ohnesorge B. Erkrankungden der Nase, Nasennebenhöhlen und Nasenmuscheln. In: Brehm W, Gehlen H, Ohnesorge B, Wehrend A, editors. Handbuch Pferdepraxis: begründet von Olof Dietz und Bernhard Huskamp. Georg Thieme Verlag; 2016. p. 327–34.

[17] Cohen NA. Sinonasal mucociliary clearance in health and disease. Ann Otol Rhinol Laryngol Suppl 2006;115:20–6.

[18] Wagenmann M, Nacclerio R. Anatomic and physiologic considerations in sinusitis. J Allergy Clin Immunol 1992;90:419–23.

[19] Messerklinger W. Über die Drainage der menschlichen Nasennebenhöhlen unter normalen und pathologischen Bedingungen. Mschr Ohrenheilk 1966;100:101.

[20] Stammberger H. Endoscopic endonasal surgery—concepts in treatment of recurring sinusitis. Part I. Anatomical and pathophysiological considerations. Otolaryngol Head Neck Surg 1986;94: 143–7.

[21] Nöller C, Nowak M, Hamann J, Fritsch G, Budras K. Klinische Anatomie der Nasen-und Nasennebenhöhlen des Pferdes-Grundlagen für die Endoskopie, Computertomographie und Chirurgie. Pferdeheilkunde 2007;23:47–58.

[22] Tatarniuk DM, Bell C, Carmalt JL. A description of the relationship between the nasomaxillary aperture and the paranasal sinus system of horses. Vet J 2010;186:216–20.

[23] Probst A, Henninger W, Willmann M. Communications of normal nasal and paranasal cavities in computed tomography of horses. Vet Radiol Ultrasound 2005;46:44–8.

[24] Nickel R, Wilkens H. Zur Topographie der Nasenhöhle und der Nasennebenhöhlen beim Pferd. Dtsch Tierarztl Wochenschr 1958; 7:173–80.

[25] Brinkschulte M, Bienert-Zeit A, Lüpke M, Hellige M, Ohnesorge B, Staszyk C. The sinonasal communication in the horse: examinations using computerized three-dimensional reformatted renderings of computed-tomography datasets. BMC Vet Res 2014;10:72.

[26] Kaminsky J, Bienert-Zeit A, Hellige M, Rohn K, Ohnesorge B. Comparison of image quality and in vivo appearance of the normal equine nasal cavities and paranasal sinuses in computed tomography and high field (3.0 T) magnetic resonance imaging. BMC Vet Res 2016;12:13.

[27] De Zani D, Borgonovo S, Biggi M, Vignati S, Scandella M, Lazzaretti S, et al. Topographic comparative study of paranasal sinuses in adult horses by computed tomography, sinuscopy, and sectional anatomy. Vet Res Commun 2010;34(Suppl 1):S13–6.

[28] Schumacher J, Honnas C, Smith B. Paranasal sinusitis complicated by inspissated exudate in the ventral conchal sinus. Vet Surg 1987; 16:373–7.

[29] Schumacher J, Crossland L. Removal of inspissated purulent exudate from the ventral conchal sinus of three standing horses. J Am Vet Med Assoc 1994;205:1312–4.

[30] Dixon PM, Froydenlund T, Luiti T, Kane-Smyth J, Horbal A, Reardon RJ. Empyema of the nasal conchal bulla as a cause of chronic unilateral nasal discharge in the horse: 10 cases (2013-2014). Equine Vet J 2015;47:445–9.

[31] Nickels FA. Nasal Passages and Paranasal Sinuses. In: Auer JA, Stick JA, editors. Equine surgery. WB Saunders; 2012. p. 557–68.

[32] Feige K, Geissbuhler U, Furst A, Ehrat F, Schwarzwald C. Sinusitis beim Pferd. Eine retprospektive Untersuchung anhand von 55 Fallen. Pferdeheilkunde 2000;16:495–501.

[33] Tremaine W, Freeman DE. Disorders of the paranasal sinuses. Equine respiratory medicine and surgery. Philadelphia: Saunders Elsevier; 2007. p. 403–4.

[34] Bienert A. Digitalradiographische, computertomographische und mikrobiologische Untersuchungen bei Backenzahnerkrankungen des Pferdes. Hannover: Tierärztliche Hochsch; 2002. Diss.

[35] Barakzai S. Rhinotomy to access lesions within the nasal cavity of two horses. Equine Vet Educ 2017;29:61–4.

[36] Zwick T. Einseitiger Nasenausfluss beim Pferd. Pferdespiegel 2013; 16:65–73.

[37] Wood J, Newton J, Chanter N, Mumford J. Inflammatory airway disease, nasal discharge and respiratory infections in young British racehorses. Equine Vet J 2005;37:236–42.

[38] Rosenfeld RM, Andes D, Bhattacharyya N, Cheung D, Eisenberg S, Ganiats TG, et al. Clinical practice guideline: adult sinusitis. Otolaryngol Head Neck Surg 2007;137:S1–31.

[39] Fokkens W, Lund V, Mullol J. European position paper on rhinosinusitis and nasal polyps 2007. Rhinol Suppl 2006;20:1–136.

[40] Meltzer EO, Hamilos DL, Hadley JA, Lanza DC, Marple BF, Nicklas RA, et al. Rhinosinusitis: establishing definitions for clinical research and patient care. J Allergy Clin Immunol 2004;114: 155–212.

[41] Scadding G, Durham S, Mirakian R, Jones N, Drake-Lee A, Ryan D, et al. BSACI guidelines for the management of rhinosinusitis and nasal polyposis. Clin Exp Allergy 2008;38:260–75.

[42] Fokkens WJ, Lund VJ, Mullol J, Bachert C, Alobid I, Baroody F, et al. European position paper on rhinosinusitis and nasal polyps 2012. Rhinol Suppl 2012;50:1–298.

[43] Slavin RG, Spector SL, Bernstein IL, Workgroup SU, Kaliner MA, Kennedy DW, et al. The diagnosis and management of sinusitis: a practice parameter update. J Allergy Clin Immunol 2005;116:S13–47.

[44] Meltzer EO, Hamilos DL. Rhinosinusitis diagnosis and management for the clinician: a synopsis of recent consensus guidelines. Mayo Clin Proc 2011;85:427–43.

[45] Anand VK. Epidemiology and economic impact of rhinosinusitis. Ann Otol Rhinol Laryngol 2004;113:3–5.

[46] Pediatrics AAO. Clinical practice guideline: management of sinusitis. Pediatrics 2001;108:798.

[47] Brook I. Sinusitis of odontogenic origin. Otolaryngol Head Neck Surg 2006;135:349–55.

[48] Taschieri S, Torretta S, Corbella S, Del Fabbro M, Francetti L, Lolato A, et al. Pathophysiology of sinusitis of odontogenic origin. J Investig Clin Dent 2017;8:e12202.

[49] Troeltzsch M, Pache C, Troeltzsch M, Kaeppler G, Ehrenfeld M, Otto S, et al. Etiology and clinical characteristics of symptomatic unilateral maxillary sinusitis: a review of 174 cases. J Craniomaxillofac Surg 2015;43:1522–9.

[50] Zirk M, Dreiseidler T, Pohl M, Rothamel D, Buller J, Peters F, et al. Odontogenic sinusitis maxillaris: a retrospective study of 121 cases with surgical intervention. J Craniomaxillofac Surg 2017;45:520–5.

[51] Benitah N. Canine nasal aspergillosis. Clin Tech Small Anim Pract 2006;21:82–8.

[52] Kuehn NF. Chronic rhinitis in cats. Clin Tech Small Anim Pract 2006;21:69–75.

[53] Crocker C, Rings D. Lymphosarcoma of the frontal sinus and nasal passage in a cow. J Am Vet Med Assoc 1998;213:1472–4.

[54] Mundell L, Smith B, Hoffman R. Maxillary sinus cysts in two cattle. J Am Vet Med Assoc 1996;209:127–9.

[55] Ward J, Rebhun W. Chronic frontal sinusitis in dairy cattle: 12 cases (1978-1989). J Am Vet Med Assoc 1992;201:326–8.

[56] Crilly JP, Hopker A. Case report: diagnosis and treatment of maxillary sinusitis in an Aberdeen Angus cow. Livestock 2013;18:230–5.

[57] Nowak M, Schulte-Bahrenberg S. Dentogene Sinuserkrankungen. in Lehrbuch der Zahnheilkunde beim Pferd. Vogt C Schattauer Verlag; 2011. p. 247–56.

[58] Barakzai S. Standing sinus surgery in the horse: indications, techniques and complications. In Practice 2008;30:252–62.

[59] Administration FDA. Guidance for industry acute bacterial sinusitis: developing drugs for treatment; 2012. https://wwwfdagov/downloads/Drugs/GuidanceComplianceRegulatoryInformation/Guidances/ucm070939pdf. Accessed August 1, 2017.

[60] Benninger MS, Appelbaum PC, Denneny JC, Osguthorpe DJ, Stankiewicz JA. Maxillary sinus puncture and culture in the diagnosis of acute rhinosinusitis: the case for pursuing alternative culture methods. Otolaryngol Head Neck Surg 2002;127:7–12.

[61] Gold SM, Tami TA. Role of middle meatus aspiration culture in the diagnosis of chronic sinusitis. Laryngoscope 1997;107:1586–9.

[62] Orobello PW, Park RI, Belcher LJ, Eggleston P, Lederman HM, Banks JR, et al. Microbiology of chronic sinusitis in children. Arch Otolaryngol Head Neck Surg 1991;117:980–3.

[63] Ferguson B, Straka M. Endoscopically directed middle meatal aspirate compared to antral tap in the diagnosis of acute and chronic bacterial rhinosinusitis. Annual Meeting of the American Rhinologic Society, New Orleans, LA; 1999.

[64] Benninger MS, Payne SC, Ferguson BJ, Hadley JA, Ahmad N. Endoscopically directed middle meatal cultures versus maxillary sinus taps in acute bacterial maxillary rhinosinusitis: a meta-analysis. Otolaryngol Head Neck Surg 2006;134:3–9.

[65] Bailey G, Love DN. Oral associated bacterial infection in horses: studies on the normal anaerobic flora from the pharyngeal tonsillar surface and its association with lower respiratory tract and paraoral infections. Vet Microbiol 1991;26:367–79.

[66] Kennedy R, Lappin DF, Dixon PM, Buijs MJ, Zaura E, Crielaard W, et al. The microbiome associated with equine periodontitis and oral health. Vet Res 2016;47:49.

[67] Kern I, Bartmann C, Verspohl J, Rohde J, Bienert-Zeit A. Bacteraemia before, during and after tooth extraction in horses in the absence of antimicrobial administration. Equine Vet J 2016;49:178–82.

[68] Bienert A, Bartmann C, Verspohl J, Deegen E. Bakteriologische Befunde bei endodontischen und apikalen Erkrankungen der Backenzähne des Pferdes. Dtsch Tierarztl Wochenschr 2003;110:358–61.

[69] Capra GG, Carbone PN, Mullin DP. Paranasal sinus mucocele. Head Neck Pathol 2012;6:369–72.

[70] Watt D. A case of cryptococcal granuloma in the nasal cavity of a horse. Aust Vet J 1970;46:493–4.

[71] Roberts M, Sutton R, Lovell D. A protracted case of cryptococcal nasal granuloma in a stallion. Aust Vet J 1981;57:287–91.

[72] Corrier D, Wilson S, Scrutchfield W. Equine cryptococcal rhinitis. Compend Contin Educ Vet 1984;6:556–8.

[73] Diseases of the Nasal Cavity and Paranasal Sinuses. In: Rush B, Mair T, editors. Equine respiratory diseases. John Wiley and Sons; 2008. p. 41–55.

[74] Freeman DE. Sinus disease. Vet Clin North Am Equine Pract 2003;19:209–43.

[75] Brook I. Aerobic and anaerobic bacterial flora of normal maxillary sinuses. Laryngoscope 1981;91:372–6.

[76] Jiang R-S, Liang K-L, Jang J-W, Hsu C-Y. Bacteriology of endoscopically normal maxillary sinuses. J Laryngol Otol 1999;113:825–8.

[77] Sobin J, Engquist S, Nord CE. Bacteriology of the maxillary sinus in healthy volunteers. Scand J Infect Dis 1992;24:633–5.

[78] Hamad WA, Matar N, Elias M, Nasr M, Sarkis-Karam D, Hokayem N, et al. Bacterial flora in normal adult maxillary sinuses. Am J Rhinol Allergy 2009;23:261–3.

[79] Albu S, Florian IS. Bacteriology of normal frontal sinuses. Am J Otolaryngol 2013;34:327–30.

[80] Kremer B, Jacobs JA, Soudijn ER. Clinical value of bacteriological examinations of nasal and paranasal mucosa in patients with chronic sinusitis. Eur Arch Otorhinolaryngol 2001;258:220–5.

[81] Van Cauwenberge P, Vander Mijnsbrugge AM, Ingels KJ. The microbiology of acute and chronic sinusitis and otitis media: a review. Eur Arch Otorhinolaryngol 1993;250:S3–6.

[82] Penttilä M, Savolainen S, Kiukaanniemi H, Forsblom B, Jousimies-Somer H. Bacterial findings in acute maxillary sinusitis—European study. Acta Otolaryngol Suppl 1997;117:165–8.

[83] Sande MA, Gwaltney JM. Acute community-acquired bacterial sinusitis: continuing challenges and current management. Clin Infect Dis 2004;39:S151–8.

[84] Hansen JG, Højbjerg T, Rosborg J. Symptoms and signs in culture-proven acute maxillary sinusitis in a general practice population. APMIS 2009;117:724–9.

[85] Mantovani K, Bisanha AA, Demarco RC, Tamashiro E, Martinez R, Anselmo-Lima WT. Maxillary sinuses microbiology from patients with chronic rhinosinusitis. Braz J Otorhinolaryngol 2010;76:548–51.

[86] Tantilipikorn P, Fritz M, Tanabodee J, Lanza DC, Kennedy DW. A comparison of endoscopic culture techniques for chronic rhinosinusitis. Am J Rhinol 2002;16:255–60.

[87] Brook I, Frazier EH. Correlation between microbiology and previous sinus surgery in patients with chronic maxillary sinusitis. Ann Otol Rhinol Laryngol 2001;110:148–51.

[88] Rombaux P, Gigi J, Hamoir M, Eloy P, Bertrand B. Bacteriology of chronic sinusitis: the bulla ethmoidalis content. Rhinology 2002;40:18–23.

[89] Doyle PW, Woodham J. Evaluation of the microbiology of chronic ethmoid sinusitis. J Clin Microbiol 1991;29:2396–400.

[90] Brook I. Bacteriology of chronic maxillary sinusitis in adults. Ann Otol Rhinol Laryngol 1989;98:426–8.

[91] Araujo E, Dall C, Cantarelli V, Pereira A, Mariante AR. Microbiology of middle meatus in chronic rhinosinusitis. Braz J Otorhinolaryngol 2007;73:549–55.

[92] Mehra P, Jeong D. Maxillary sinusitis of odontogenic origin. Curr Allergy Asthma Rep 2009;9:238–43.

[93] Brook I. Microbiology of acute and chronic maxillary sinusitis associated with an odontogenic origin. Laryngoscope 2005;115:823–5.

[94] Tress B, Dorn ES, Suchodolski JS, Nisar T, Ravindran P, Weber K, et al. Bacterial microbiome of the nose of healthy dogs and dogs with nasal disease. PloS One 2017;12:e0176736.

[95] Norsworthy G. Surgical treatment of chronic nasal discharge in 17 cats. J Vet Med (USA) 1993;6:526–37.

[96] Michiels L, Day M, Snaps F, Hansen P, Clercx C. A retrospective study of non-specific rhinitis in 22 cats and the value of nasal cytology and histopathology. J Feline Med Surg 2003;5:279–85.

[97] Johnson LR, Foley JE, De Cock HE, Clarke HE, Maggs DJ. Assessment of infectious organisms associated with chronic rhinosinusitis in cats. J Am Vet Med Assoc 2005;227:579–85.

[98] Dorn ES, Tress B, Suchodolski JS, Nisar T, Ravindran P, Weber K, et al. Bacterial microbiome in the nose of healthy cats and in cats with nasal disease. PloS One 2017;12:e0180299.

[99] Murray GM, O'Neill RG, Lee AM, McElroy MC, More SJ, Monagle A, et al. The bovine paranasal sinuses: bacterial flora, epithelial expression of nitric oxide and potential role in the in-herd persistence of respiratory disease pathogens. PloS One 2017;12:e0173845.

[100] Stephenson M-F, Mfuna L, Dowd SE, Wolcott RD, Barbeau J, Poisson M, et al. Molecular characterization of the polymicrobial flora in chronic rhinosinusitis. J Otolaryngol Head Neck Surg 2010;39:182–7.

[101] Vickery TW, Kofonow JM, Ramakrishnan VR. Characterization of sinus Microbiota by 16S sequencing from swabs. Methods Mol Biol 2017;1616:23.

[102] Wilson MT, Hamilos DL. The nasal and sinus microbiome in health and disease. Curr Allergy Asthma Rep 2014;14:485.

[103] Feazel LM, Robertson CE, Ramakrishnan VR, Frank DN. Microbiome complexity and Staphylococcus aureus in chronic rhinosinusitis. Laryngoscope 2012;122:467–72.
[104] Mansmann RA, Mansmann JA. Cytology of equine nasal secretions. J Am Vet Med Assoc 1969;154:1037–42.
[105] Goldwyn BG, Sakr W, Marks SC. Histopathologic analysis of chronic sinusitis. Am J Rhinol 1995;9:27–30.
[106] Lane J, Longstaffe J, GIBBS C. Equine paranasal sinus cysts: a report of 15 cases. Equine Vet J 1987;19:537–44.
[107] Dixon P. Equine maxillary cysts. Vet Clin North Am Equine Pract 1985;7:25–33.
[108] Wilson NW, Jalowayski AA, Hamburger RN. A comparison of nasal cytology with sinus x-rays for the diagnosis of sinusitis. Am J Rhinol 1988;2:55–9.
[109] Gill FF, Neiburger JB. The role of nasal cytology in the diagnosis of chronic sinusitis. Am J Rhinol 1989;3:13–5.
[110] Druce HM. Diagnosis of sinusitis in adults: history, physical examination, nasal cytology, echo, and rhinoscope. J Allergy Clin Immunol 1992;90:436–41.
[111] Bhattacharyya N. The cytology and microbiology of persistent paranasal sinus secretions after endoscopic sinus surgery: a controlled study. Laryngoscope 2007;117:2041–4.
[112] Gupta N, Kaur J, Srinivasan R, Das A, Mohindra S, Rajwanshi A, et al. Fine needle aspiration cytology in lesions of the nose, nasal cavity and paranasal sinuses. Acta Cytol 2011;55:135–41.
[113] Cohen M, Bohling MW, Wright JC, Welles EA, Spano JS. Evaluation of sensitivity and specificity of cytologic examination: 269 cases (1999–2000). J Am Vet Med Assoc 2003;222:964–7.
[114] Ainsworth D, Cheetham J. Disorders of the respiratory system. In: Reed S, Bayly WM, Sellon DC, editors. Equine internal medicine. 3rd ed. Elsevier Health Sciences; 2009. p. 290–371.
[115] May A, Gehlen H. Durchführung, Analyse und Aussagekraft von Tracheobronchialsekret (TBS) und Bronchoalveolärer Lavage (BAL) bei Pferden mit Lungenerkrankungen. Pferdeheilkunde 2009;25:310–20.
[116] Dieckmann M, Deegen E. Klinische Bedeutung der Tracheobronchialsekret-Zytologie. Pferdeheilkunde 1990;6:101–10.

3 Ergebnisse

3.1 Publikation II

A prospective study on the microbiological examination of secretions from the paranasal sinuses in horses in health and disease

Published in: Acta Veterinaria Scandinavica
Volume 60, Issue 1, 2018, Pages 43–51

Received: 13 February 2018
Accepted: 23 June 2018
Available online: 5 July 2018

H. Gergeleit[1], J. Verspohl[2], J. Rohde[2], Carl Rohn[3], B. Ohnesorge[1†], A. Bienert-Zeit[1†]
[1] Clinic for Horses, [2] Institute for Microbiology, [3] Department of Biometry, Epidemiology and Information Processing,
University of Veterinary Medicine Hannover, Foundation, Hannover, Germany

[†] = equal contributors

Abstract

Background: Diagnostics in equine sinusitis can be challenging and often require a combination of different imaging tools to ascertain its underlying aetiology. The bacterial flora of healthy and diseased paranasal sinuses, respectively, has only sporadically been assessed in horses. The objectives of this study were to determine whether assessment of microbiological features of secretions from the paranasal sinuses displays a useful diagnostic tool in equine sinusitis to distinguish between different aetiologies. Secretion samples from 50 horses with sinusitis and from 10 healthy horses were taken transendoscopically from the drainage angle of the nasomaxillary aperture using a guidable Swing Tip catheter. Bacteria found in healthy and diseased equine sinuses were compared. Endoscopic samples in all healthy and 19 diseased horses were compared with samples taken directly from the affected sinus after trephination.

Results: Eleven of the 14 horses with primary sinusitis revealed growth of *Streptococcus equi* ssp. *zooepidemicus*, with three samples yielding pure cultures. Anaerobes were found in 15 out of 26 samples from horses with dental sinusitis. Healthy sinuses revealed mainly α-haemolytic streptococci and coagulase-negative staphylococci or showed no growth. *Enterobacteriaceae* were found more frequently in secondary sinusitis. There were significant differences in the bacterial composition and diversity ($P < 0.05$) between primary sinusitis, dental sinusitis and healthy controls. The correlation between endoscopic and trephination samples was satisfying.

Conclusions: Microbiological examination of secretions from horses with sinusitis collected transendoscopically can help to distinguish between primary and dental sinusitis. Therefore, it may display a feasible ancillary diagnostic tool, but does not replace a meticulous examination procedure including diagnostic imaging.

Gergeleit *et al. Acta Vet Scand (2018) 60:43
https://doi.org/10.1186/s13028-018-0394-4

Acta Veterinaria Scandinavica

RESEARCH Open Access

A prospective study on the microbiological examination of secretions from the paranasal sinuses in horses in health and disease

Hauke Gergeleit[1*], Jutta Verspohl[2], Judith Rohde[2], Karl Rohn[3], Bernhard Ohnesorge[1†] and Astrid Bienert-Zeit[1†]

Abstract

Background: Diagnostics in equine sinusitis can be challenging and often require a combination of different imaging tools to ascertain its underlying aetiology. The bacterial flora of healthy and diseased paranasal sinuses, respectively, has only sporadically been assessed in horses. The objectives of this study were to determine whether assessment of microbiological features of secretions from the paranasal sinuses displays a useful diagnostic tool in equine sinusitis to distinguish between different aetiologies. Secretion samples from 50 horses with sinusitis and from 10 healthy horses were taken transendoscopically from the drainage angle of the nasomaxillary aperture using a guidable Swing Tip catheter. Bacteria found in healthy and diseased equine sinuses were compared. Endoscopic samples in all healthy and 19 diseased horses were compared with samples taken directly from the affected sinus after trephination.

Results: Eleven of the 14 horses with primary sinusitis revealed growth of *Streptococcus equi* ssp. *zooepidemicus*, with three samples yielding pure cultures. Anaerobes were found in 15 out of 26 samples from horses with dental sinusitis. Healthy sinuses revealed mainly α-haemolytic streptococci and coagulase-negative staphylococci or showed no growth. *Enterobacteriaceae* were found more frequently in secondary sinusitis. There were significant differences in the bacterial composition and diversity ($P < 0.05$) between primary sinusitis, dental sinusitis and healthy controls. The correlation between endoscopic and trephination samples was satisfying.

Conclusions: Microbiological examination of secretions from horses with sinusitis collected transendoscopically can help to distinguish between primary and dental sinusitis. Therefore, it may display a feasible ancillary diagnostic tool, but does not replace a meticulous examination procedure including diagnostic imaging.

Keywords: Anaerobes, Endoscopy, Horse, Microbiology, Sinusitis, *Streptococcus equi* ssp. *zooepidemicus*

Background

Sinusitis is a common and well described disorder in horses [1–3]. However, distinguishing between a primary sinusitis and those secondary to other disorders can be challenging. Although the diagnosis of sinusitis is usually easily made by assessment of clinical signs and endoscopic findings, an exact aetiological diagnosis can often only be achieved by a combination of several imaging diagnostic tools [3, 4]. In general practice, horses with unilateral nasal discharge are usually treated with antibiotics over varying periods of time and often without permanent remission of clinical symptoms [1]. This can

*Correspondence: hauke.gergeleit@tiho-hannover.de
†Bernhard Ohnesorge and Astrid Bienert-Zeit contributed equally to this work
[1] Clinic for Horses, University of Veterinary Medicine Hannover, Foundation, Bünteweg 9, 30559 Hannover, Germany
Full list of author information is available at the end of the article

Table 1 Distribution of the etiology of sinusitis

Aetiology	Number of cases
Dental sinusitis	26
Primary sinusitis	14
Paranasal sinus cyst	2
Progressive ethmoidal haematoma	2
Traumatic sinusitis	2
Conchal necrosis	1
Malignant neoplasia	3
Total	50

make the healing process frustrating for the owner and the treating veterinarian.

The predominant organisms found in acute sinusitis in human medicine are *Haemophilus influenzae, Streptococcus pneumoniae* and *Moraxella catarrhalis*. A greater variation of bacteriological cultures exists in chronic sinusitis [5]. Maxillary sinusitis with an odontogenic origin in humans has been proven to be dominated by Gram-negative anaerobic bacteria [6]. The role of anaerobes has also been assessed for periodontitis, and endodontic and apical tooth infections in horses [7, 8], with reports of their involvement in equine dental sinusitis [9]. Nevertheless, the microbiological findings of secretions from the paranasal sinuses in horses have only been reported sporadically and have not yet been subject to wide case series or prospective studies [10].

The aims of this study were to assess whether the microbiological features of secretions from horses affected by sinusitis allow the drawing of conclusions regarding the underlying etiology and, therefore, whether this approach displays an ancillary diagnostic tool in equine sinusitis.

Methods

Patient group

The patient group (Table 1) consisted of 50 horses, including 39 Warmbloods, 5 ponies, 3 Icelandic horses, 1 English Thoroughbred, 1 Arabian Thoroughbred and 1 Coldblood (25 mares, 23 geldings and 2 stallions) with a mean ± SD age of 12.1 ± 6.3 years (range 1–27 years). They were presented to the Clinic for Horses of the University of Veterinary Medicine Hannover, Foundation, with a history of unilateral nasal discharge between October 2016 and September 2017. All horses were diagnosed with a paranasal sinusitis. Their history included the duration of clinical signs, previous medical and surgical treatments, and response to those treatments prior to referral. All horses underwent a detailed clinical examination and a specific clinical examination of the head.

Transnasal endoscopy and sampling

The endoscopic examination of the respiratory tract was performed using an Olympus video endoscope (PCF-H180AL).[1] All horses were restrained in stocks and sedated using detomidine (Cepesedan,[2] 0.01 mg/kg bwt i.v.) combined with butorphanol (Butorgesic,[3] 0.01 mg/kg bwt i.v.).

Particular attention was paid to the nasal conchae, the nasal meati, with special regard to the caudal aspect of the middle nasal meatus (i.d. 'drainage angle'), and the ethmoid region. Presence, quantity and nature of any secretion draining off the drainage angle were noted (Fig. 1). The endoscope was positioned in the middle nasal meatus rostral to the draining track. A flexible, wire-guided catheter (Olympus Swing Tip$_{TM}$ PR-233Q[4] [11]) was advanced through the working channel of the endoscope. Its distal tip was angulated using the handle attached until the catheter was positioned in the drainage angle and secretion was aspirated into the catheter. The catheter was removed before endoscopic examination was continued. The same method was used on both sides of the head in 35 horses to compare the findings between the usually unilaterally occurring symptoms in one nasal cavity and the non-affected side for an intra individual comparison. Airway endoscopy was completed to exclude other diseases.

Depending on history, clinical signs and findings during the endoscopic examination, further diagnostic tools including examination of the oral cavity, radiography and computed tomography of the head were combined to make an etiological diagnosis.

Nineteen horses underwent trephination or bone flap surgery of the diseased sinus for therapeutic reasons. In these cases a direct swab sample of the secretion or mass was taken for microbiological examination and compared to the sample collected transendoscopically to assess the method's accuracy. The correlation between both samples was stated as good when they were dominated by the same microorganisms.

Control group

The control group consisted of 10 horses, including 6 Warmbloods, 3 Trotters and 1 Icelandic horse (1 gelding and 9 mares) with a mean ± SD age of 16.6 ± 7.3 years (range 2–30 years) that were owned by the Clinic for Horses of the University of Veterinary Medicine Hannover, Foundation. The horses had no known history of sinonasal diseases and had not received antimicrobial drugs for at least 4 weeks prior to the study. They underwent a

[1] Olympus Deutschland GmbH, Hamburg, Germany.

[2] CP-Pharma Handelsgesellschaft mbH, Burgdorf, Germany.

[3] CP-Pharma Handelsgesellschaft mbH, Burgdorf, Germany.

[4] Olympus Deutschland GmbH, Hamburg, Germany.

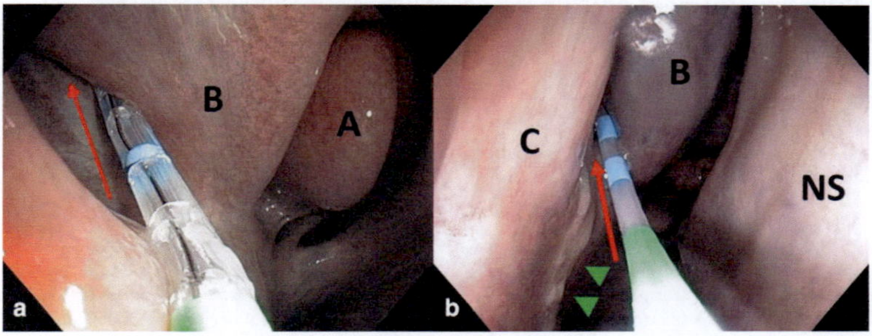

Fig. 1 Endoscopic images of the right 'drainage angle' (red arrow) with the Swing Tip-catheter in place. The catheter is advanced along the drainage angle and secretions are aspirated. a) In a healthy control horse. b) In a horse with squamous cell carcinoma of the right sinus system with purulent, malodorous discharge (green arrowheads). A: middle nasal concha, B: dorsal nasal concha, C: ventral nasal concha, NS: nasal septum

detailed clinical and radiographic examination of the head. The horses were put under general anesthesia for an unrelated terminal study. All procedures were approved by the Lower Saxony State Office for Consumer Protection and Food Safety (reference no. 33.19-42502-04-16/2212). Signs of sinonasal disorders were excluded during endoscopic examination before secretion samples were aspirated using the technique described previously. Afterwards, the ipsilateral frontal sinus was trephined under sterile conditions and a swab was inserted through the frontomaxillary opening to collect a mucosal smear from the caudal maxillary sinus for microbiological examination.

Microbiological processing

The secretion specimens were applied to sterile cotton swabs that were placed into a semi-solid Amies transport medium[5] and transferred to the Institute for Microbiology at the University of Veterinary Medicine Hannover, Foundation. The swabs were streaked on different solid media and incubated under aerobic, microaerophilic and anaerobic conditions. After streaking on solid media swabs were placed in enrichment broth which was incubated aerobically overnight and then plated on solid media. Further differentiation was achieved by following the Institute's instructions for processing and examination. The degree of bacterial growth was assessed semi-quantitatively.

Measures to decrease the risk of contamination

Precautions were taken to minimize the risk of contamination during transendoscopic sampling. In those horses that underwent transendoscopic sampling of both sides the sample from the clinically inconspicuous side was always taken first. We assumed lower bacterial burden for the control side and therefore minor risk of contamination on to the diseased side. The catheter with the sample was removed and the diseased side was sampled with a new, sterile catheter. Sterile gloves were used when handling the catheter. After each examination the endoscope and the catheters were thoroughly cleaned and disinfected according to the standards of the Clinic for Horses before they were used again. Each catheter was used approximately ten times before it was disposed.

Statistical analysis

Statistical analysis was performed using the statistic software SAS, Version 9.4.[6] Distinction of bacterial profiles between primary sinusitis, dental sinusitis and healthy controls was tested using Chi squared test. Fisher's exact test was used to compare the prevalence of certain bacterial species and families between healthy and diseased sinuses. Wilcoxon–Mann–Whitney test was used to compare diversity between primary and secondary dental sinusitis as well as between healthy and diseased sinuses. P-values < 0.05 were considered statistically significant.

Results

It was possible to place the catheter precisely in the drainage angle of the nasomaxillary aperture in all 60 horses. The amount and character of the secretions varied and secretions were more difficult to aspirate when they were of low quantity or high viscosity. The most

[5] Meus S.r.l, Piove di Sacco, Italy.

[6] SAS Institute, Cary, NC, USA.

Table 2 Distribution of certain bacterial species among primary sinusitis, dental sinusitis and healthy sinuses

Primary sinusitis	n	*S. zooepidemicus*			Anaerobes			*Enterobacteriaceae*		
	14	11	78.6%	A*, B**	1	7.1%		2	14.3%	
Dental sinusitis	26	7	26.9%		15	57.7%	B**, C*	12	46.2%	B
Healthy sinuses	45	3	6.7%		0	0%		8	17.8%	

A: significant compared against dental sinusitis; B: significant compared against healthy sinuses; C: significant compared against primary sinusitis

* ($P<0.01$); ** ($P<0.001$)

common cause of sinusitis was dental disease (26 horses), followed by primary sinusitis (14 horses). Less common causes included neoplasia, sinus cysts, progressive ethmoidal haematoma (PEH), trauma and conchal necrosis (Table 1). Samples from healthy sinuses included the horses from the control group (n = 10) and from the contralateral, non-affected side in 35 horses from the diseased group. The bacterial profiles of horses with primary sinusitis, dental sinusitis and healthy controls were significantly distinct ($P<0.01$).

Primary sinusitis

The culture rate for horses with primary sinusitis was 100% with a median of two bacterial species per specimen (Min 1, Max 6, Range 5). *Streptococcus equi* ssp. *zooepidemicus* (*S. zooepidemicus*) was isolated transendoscopically in eleven horses, revealing pure growths in three cases and showing moderate to heavy growth in ten samples. *S. zooepidemicus* was isolated significantly more often in primary sinusitis than in dental sinusitis ($P<0.01$) and healthy sinuses ($P<0.001$) (Table 2). Moderate and high numbers of *S. zooepidemicus* were encountered significantly more often in primary than in dental sinusitis ($P<0.001$). Eight horses had additional growth of Gram-negative bacteria, including two horses revealing *Enterobacteriaceae* and one horse with a heavy growth of *Pseudomonas aeruginosa*. One horse had a growth of mixed Gram-negative bacteria only. The antibiogram of the latter two horses showed the bacterial isolates to be resistant to the antibiotics used previously. Anaerobes were detected in only one horse with acute onset of primary sinusitis 3 days after undergoing a median laparotomy. There was a good correlation between the microbiological findings of endoscopic and trephination specimens (Table 4).

Dental sinusitis

The culture rate for horses with dental sinusitis was 100% for the samples collected transendoscopically. All samples revealed a median of three bacterial species (Min 1, Max 9, Range 8). Sinusitis with a dental origin showed a significantly higher diversity in the microbiological profile than those in primary sinusitis ($P<0.05$) and healthy sinuses ($P<0.001$). Fifteen samples showed mixed bacterial growth of facultatively and strictly anaerobic bacteria of which 13 showed growth of more than one anaerobic species. The frequent occurrence of obligate anaerobes is a significant difference to primary sinusitis ($P<0.01$) and healthy controls ($P<0.001$). Genera were dominated by *Fusobacterium* spp. and *Porpyhromonas* spp., and also included *Bacteroides* spp., *Prevotella* spp., Gram-negative and -positive obligatory anaerobes that could not be identified to the genus level, and *Peptostreptococcus* spp. (Table 3). The prevalence of anaerobic bacteria rises to 18 horses when horses in which anaerobes were detected from trephination samples are included.

There was a higher prevalence of *Enterobacteriaceae* in horses with dental sinusitis (46.2%) than in those with primary sinusitis (14.3%), and a significantly higher prevalence than in healthy controls ($P<0.05$), with *Escherichia coli* being encountered the most frequently. *S. zooepidemicus* was present in seven samples, with only two of them revealing moderate to high numbers.

Other forms of sinusitis

Sinusitis that had an aetiology other than primary or odontogenic was rare in the current study. *S. zooepidemicus* was present in two cases of malignant neoplasia

Table 3 Number of strictly anaerobic isolates in dental sinusitis (n = 15)

Bacteroides spp.	1
Bacteroides caccae	1
Bacteroides pyogenes	2
Fusobacterium spp.	9
Fusobacterium equinum	1
Fusobacterium necrophorum	2
Fusobacterium varium	1
Gram-negative anaerobes	5
Gram-positive anaerobes	1
Peptostreptococcus spp.	2
Prevotella spp.	4
Porphyromonas spp.	6
Total	35

Table 4 Bacterial isolates from transendoscopic samples and swabs taken directly after trephination

Bacterial isolate	Primary				Dental													Cyst	PEH
	Horse no.																		
	1	2	3	4	5	6	7	8	9	10	11	12	13	14	15	16	17	18	19
Aerobic and facultatively anaerobic bacteria																			
Acinetobacter baumannii							+												
Acinetobacter calcoaceticus				O															
Acinetobacter johnsonii	+																		
Actinobacillus equuli ssp. *haemolyticus*				O	T								T						
Actinobacillus lignieresii												O							
Alcaligenes faecalis						T													
Bacillus spp.									+					+					
Bordetella spp.			+					O							O				
Citrobacter braakii						T													
Coryneforme bacteria											O				O				
Cronobacter sakazakii								+											
Enterococcus faecalis					T														
Escherichia coli							+					+					+		
Gram-positive bacteria*										T									
Klebsiella oxytoca																			+
Morganella morganii					T														
Non-fermenter				+															
Pantoea agglomerans													+						
Pantoea spp.								+											
Pseudomonas aeruginosa		O					T		T				+			+			
Pseudomonas spp.						T										+			
Streptococcus equi ssp. *zooepidemicus*	O	+	O		O							O	O	+				O	
Serratia liquefaciens						T		O											
Staphylococci (coagulase-negative)			+	T		+	T		O					T	+			+	
Staphylococcus succinus										+									
Staphylococcus xylosus										+									
Streptococci (alpha-haemolytic)						+	T	O	O			O	O			O		+	
Streptomyces spp.								+											
Vagococcus fluvalis						T													
Strictly anaerobic bacteria																			
Bacteroides pyogenes					T											T	O		
Fusobacterium equinum									+			T							
Fusobacterium spp.					T	T		O			O				O		+		
Fusobacterium varium										O			+						
Gram-negative anaerobic bacteria*								+		T		T	+		O	O			
Porphyromonas spp.								O	+	O	O								
Prevotella heparinolytica												T							
Prevotella spp.									O	O			T						

+, organism isolated from endoscopic sample; T, organism isolated from trephination sample; O, organism isolated from both: endoscopic and corresponding trephination sample; PEH, progressive ethmoidal haematoma

* Isolates could not be identified to the genus level

and in a paranasal sinus cyst, whereas a second case of a sinus cyst showed mixed facultatively anaerobic Gram-negative bacteria. One case of a squamous cell carcinoma of the right sinus system revealed growth of *Bacteroides fragilis*, *Porphyromonas* spp. and *Escherichia coli*. This horse had an extraction of 109 and 110 3 months prior to referral. Obligate anaerobic bacteria and *Enterobacteriaceae* were also present in a case of conchal necrosis. Sanguineous secretions in two cases of traumatic sinusitis due to fractures of frontal or maxillary bones revealed coagulase-negative staphylococci (CNS) in small numbers, whereas sanguineous nasal discharge in two horses in association with a PEH showed mixed facultatively anaerobic Gram-negative bacteria. The PEH itself did not yield any bacterial growth. These results cannot be analyzed statistically due to the small sample sizes.

Healthy sinuses

The overall culture rate for healthy sinuses was 73.3% (33 out of 45 samples), with all positive cultures revealing a median of two bacterial species (Min 1, Max 5, Range 4). None of these samples revealed obligate anaerobes and only five showed low to moderate numbers of *Enterobacteriaceae*. The CNS and alpha-haemolytic streptococci (53.3% of positive cases) were the genera most frequently isolated. *S. zooepidemicus* was present in low numbers in three cases that were diagnosed with a primary sinusitis of the contralateral diseased side.

Sinus surgery

Nineteen horses underwent therapeutic sinus surgery. Overall, there was a good correlation of the direct samples with the corresponding sample collected transendoscopically in 79% (Table 4). Bacteria that were estimated to play a dominant role in the endoscopic sample could also be verified in the direct swab sample in 15 out of 19 sample pairs. The endoscopic sample failed to reveal growth of anaerobic bacteria in comparison to the direct sample in three cases of dental sinusitis, whereas two other cases did not reveal anaerobic bacteria in either of the samples. There was good correlation of the endoscopic and the direct sample in all four horses with primary sinusitis. One horse with a paranasal sinus cyst was positive for *S. zooepidemicus* in both samples.

Samples that were taken from the caudal maxillary sinus after trephination in the control group did not reveal bacterial growth in six cases and low numbers of Gram-positive bacteria in four cases, whereas the corresponding transendoscopic samples showed mixed facultatively anaerobic Gram-negative and -positive bacteria in four horses and low numbers of Gram-positive bacteria and no bacterial growth in three horses each.

Discussion

Diagnostics and successful treatment of sinusitis in horses can be challenging and frustrating to both the owner and the treating veterinarian in cases of prolonged and recurrent courses of this frequently described disease [3, 4]. Almost all cases included in this study had a history of chronic or recurrent nasal discharge of up to 5 years without permanent remission of symptoms with antibiotic treatment.

Symptoms of sinusitis were easily identified during clinical and endoscopic examination with varying severity of the disease. This is the first time that transendoscopic collection of secretions from the paranasal sinuses is reported. If discharge from the nasomaxillary aperture was present, the guidable Swing Tip catheter was easily positioned without causing mucosal bleedings. The use of the handle to control the bend of the tip before advancing the catheter along the drainage angle, facilitated sampling of secretions without gross contamination by touching the nasal mucosa or mucus present in the nasal cavity. In the absence of distinct discharge, it was possible to aspirate small amounts of serous secretions from the drainage angle.

Transendoscopic sampling for bacteriological culture holds the risk of contamination either through the endoscope or the catheter, if sterility of the utensils is not achieved, or through failure in handling. In this study, samples from the endoscope or the catheters to control for bacterial contamination were not taken, thus interference with the results cannot be fully excluded. Nevertheless the results indicate that our sampling and cleaning methods are extensively safe to avoid severe contaminations. The absence of bacteria or its detection in low numbers in most of the samples from the control group allows to draw conclusions that the risk of contamination was generally low. Moreover the findings between diseased and healthy sinuses differed significantly and some samples from horses with primary sinusitis even revealed pure cultures. However, the sample from the contralateral, non-affected sinus system in two horses revealed just as heavy growth of bacteria as the affected side, suggesting possible contamination between the two sides through the nasal cavity and the nasopharyngeal region. Undetected defects in the nasal septum display less often but are also possible reasons for these results [12]. Overall, the risk of interference with the results due to possible contamination can be regarded as low. The samples were taken following the same protocol and therefore had been exposed to an equal risk of contamination. Nevertheless, the bacterial findings between primary and dental sinusitis differed significantly in their composition and diversity, which remains the essential statement.

It was possible to obtain a direct sample during sinus surgery to assess the accuracy of the transendoscopic sampling method in 19 diseased horses. There was good correlation of the sample pairs in about 80% (Table 4), but complete consistency was not achieved in any of the sample pairs, probably due to the heterogeneous nature of secretions. Additionally, in most cases, 1 or 2 days passed between the endoscopic sampling and the trephination, during which antibiotic therapy had usually already been started. In three cases, the endoscopic samples failed to reveal anaerobes, whereas the corresponding trephination samples did. In one of these horses, characteristic secretion tracks along the drainage angle were absent, albeit large amounts of pus were present within the affected sinuses. Drainage was likely to be restricted due to inspissation of pus and narrowing of the drainage pathways. In the other two horses, purulent secretions with characteristic malodour could be collected. Since anaerobes have fastidious growth requirements, failure in handling, including prolonged storing, might have resulted in negative bacteriological culture.

Streptococcus equi ssp. *zooepidemicus* was the predominant organism associated with primary sinusitis in the current study, but was also isolated in association with some odontogenic and neoplastic conditions. These findings are similar to previous descriptions of this bacterial species in single cases of primary and dental sinusitis [4, 13, 14]. There was a distinctly smaller prevalence and it was less quantitative in secondary sinusitis overall. However, its presence in both primary and secondary sinusitis outlines the importance of diagnostic imaging in equine paranasal sinus diseases to exclude dental-related conditions and space-occupying processes. Microbiological evidence of *S. zooepidemicus* in two cases of malignant neoplasia and a paranasal sinus cyst in the current study indicate that its occurrence is not only related to primary inflammation of the sinus mucosa, but also to restricted sinonasal drainage, resulting in a reduction of mucociliary clearance [15]. *S. zooepidemicus* is often referred to as a mucosal commensal of the upper respiratory tract of healthy horses [16]. Nevertheless, none of the trephined healthy sinuses revealed growth of *S. zooepidemicus*, but it was detected in three endoscopic samples from healthy sinuses. These horses were diagnosed with a primary sinusitis of the contralateral sinus system with microbiological evidence of *S. zooepidemicus*, indicating possible contamination of the control side. *S. zooepidemicus* is a frequently isolated opportunistic pathogen that is often associated with severe pneumonia in young and middle-aged horses [16, 17]. The bacterium possesses numerous virulence factors, including adhesion capacities, and exerts an anti-phagocytic activity through the presence of hyaluronic acid and immunoglobulin-binding proteins [18]. The pathogenic potential of *S. zooepidemicus* is enhanced by the presence of pre-adapted variants of the bacterium that can adapt immediately to changes in its environment, such as alterations in the host's immune status or transmission from a carrier horse to another susceptible individual [19].

Infections with *S. zooepidemicus* are not limited to the respiratory tract, but are also associated with septicaemia, placentitis, endometritis, peritonitis and arthritis, among others, in a wide range of species, including horses, livestock, dogs and camelids [20–22]. Although rarely described, transmissions from horses to humans are possible with cases of severe sepsis and meningitis, underlining the importance of recognizing this infection as a potential zoonosis [23]. The causative and highly contagious agent of strangles, *Streptococcus equi* ssp. *equi*, another closely related, but host-restricted β-haemolytic *streptococcus equi* subspecies with a much narrower tissue tropism [24], was not isolated in any of the samples of this study, although it had been described in equine sinusitis previously [4].

Enterobacteriaceae were isolated more frequently in horses with dental sinusitis than in healthy controls and in those with a primary sinusitis, although in the latter, these results did not reach statistical significance. Our findings outline parallels to human chronic sinusitis, where *Enterobacteriaceae* are considered as important infective agents with a secondary role [25].

Both aerobic and anaerobic bacteria were isolated from swab samples with a significant predominance of anaerobes in horses with dental sinusitis. The close relationship of the upper caudal cheek teeth to the maxillary sinuses makes the latter prone to becoming secondarily infected by dental diseases [2]. Obligatory anaerobic bacteria have been isolated previously from swab samples in horses affected by dental diseases [7, 8, 26] and have also been described in horses with dental sinusitis [9]. There is a good correlation between those findings and the results of this study. The microbiological profile of the horse with a conchal necrosis showed similarities to dental sinusitis, including anaerobic bacteria and *Enterobacteriaceae*. Notwithstanding, this horse had no known history of dental diseases and showed no radiological signs of a dental disease when presented, albeit malodorous unilateral nasal discharge was present indicating infection with strictly anaerobic bacteria. Historical, clinical and microbiological findings coincide with those of a case series about horses affected by conchal necrosis [27]. No anaerobes were found in the control group, either in the samples collected transendoscopically nor in the samples collected directly after trephination, suggesting that these organisms are not part of the physiological sinus flora in horses. However, they might be detectable using culture-independent broad platform molecular techniques. Results of microbiological findings in human medicine differ significantly in their diversity and composition when conventional cultivating methods are compared to molecular techniques [28].

The sole use of conventional bacterial culture methods displays a limitation of this study, since investigations in humans have shown that many bacteria cannot be cultured using conventional approaches [29]. Conventional culture methods and culture-independent analysis of the microbiological contents of middle nasal meatus specimens from humans with chronic rhinosinusitis showed broad consistency, but culture failed to detect potentially pathogenic bacteria [30]. Thus, it has been proven that the sequence-based approach provides a more precise phylogenetic resolution and detects greater biodiversity. Therefore, molecular genetic methods may also be beneficial for further investigations into the flora of equine paranasal sinuses.

The limiting factor with the biggest impact on this study's outcome is probably the inhomogeneity of the patients group, with many horses being treated previously with various antimicrobial drugs over different periods of time. Prior antibiotic therapy is likely to cause incalculable alterations in the microbiological composition of the samples. Two cases of horses in this study with a primary sinusitis that were reluctant to multiple antibacterial drug administration, underline this theory. Both horses revealed Gram-negative bacteria that were resistant to the antibiotics that had been used previously. It would be desirable to repeat this study on horses without any previous antibacterial treatments to get more precise results.

The authors tried to identify specific pathogens in the present study that could serve as indicators for different parasinuidal diseases. However, the view of bacterial involvement in rhinosinusitis in human medicine has expanded beyond that of infection with individual pathogens to considerations of increased bacterial abundance, biofilm formation, intracellular bacteria and alterations in the microbial community (i.e. "microbiome") [28]. Biofilm formation displays an important survival mechanism for microorganisms through attachment to surfaces and has been confirmed in human chronic sinusitis [31]. The extent to which bacteria in biofilms are cultivable has not been well studied in humans, and to the best of the authors' knowledge, no attempts have been made to assess it in equine medicine. These bacterial survival mechanisms might not only add further variation to the detection of bacteria in this study and in microbiological examinations in equine practice in general, but may also play an essential role in recurring sinusitis and cases refractory to conservative therapy. It underlines the importance of surgical treatments and sinus irrigations to interrupt the vicious circle of impaired sinonasal drainage and progression of the disease.

Extensive longitudinal studies starting with healthy young horses and follow-up on their secretion and microbiota of the paranasal sinuses until development of sinusitis would be required to evaluate all these variable factors of host-microbial interactions.

Conclusions

The results presented indicate that microbiological examination of secretions collected transendoscopically from horses with paranasal sinusitis can help to distinguish between primary sinusitis and those with a dental origin. Therefore, it may constitute a feasible ancillary diagnostic tool given that the sample is taken with the lowest possible risk of contamination. It is important to take the occurrence of diverse, polymicrobial cultures into consideration. However, microbiological examination does not replace a meticulous clinical examination of the head and the oral cavity and diagnostic imaging.

Authors' contributions

HG contributed to the study design, data collection, study execution, data analysis and interpretation. ABZ and BO contributed to the study design, study execution, data analysis and interpretation. JV and JR contributed to the study execution, data analysis and interpretation. KR contributed to statistical evaluations. All authors read and approved the final version of the manuscript.

Author details

[1] Clinic for Horses, University of Veterinary Medicine Hannover, Foundation, Bünteweg 9, 30559 Hannover, Germany. [2] Institute for Microbiology, University of Veterinary Medicine Hannover, Foundation, Bischofsholer Damm 15, 30173 Hannover, Germany. [3] Department of Biometry, Epidemiology and Information Processing, University of Veterinary Medicine Hannover, Foundation, Bünteweg 2, 30559 Hannover, Germany.

Acknowledgements

The authors acknowledge the contributions of all coworkers at the Clinic for Horses, University of Veterinary Medicine, Hannover, for assistance with sample acquisition and all coworkers at the Institute for Microbiology, University of Veterinary Medicine, Hannover, for sample processing.

Competing interests

The authors declare that they have no competing interests.

Availability of data and materials

The datasets used and/or analyzed during the current study are available from the corresponding author on reasonable request.

Consent for publication

Not applicable.

Ethics approval and consent to participate

The experimental protocol for the control group was approved by the Lower Saxony State Office for Consumer Protection and Food Safety (reference no. 33.19-42502-04-16/2212).

Horse-owners gave informed consent for their animals' inclusion in the study.

Funding

The study was not supported by external financiers.

Publisher's Note

Springer Nature remains neutral with regard to jurisdictional claims in published maps and institutional affiliations.

Received: 13 February 2018 Accepted: 23 June 2018
Published online: 05 July 2018

References

1. Dixon PM, Parkin TD, Collins N, Hawkes C, Townsend NB, Fisher G, et al. Historical and clinical features of 200 cases of equine sinus disease. Vet Rec. 2011;169:439.
2. O'Leary JM, Dixon PM. A review of equine paranasal sinusitis. Aetiopathogenesis, clinical signs and ancillary diagnostic techniques. Equine Vet Educ. 2011;23:148–59.
3. Beard W, Hardy J. Diagnosis of conditions of the paranasal sinuses in the horse. Equine Vet Educ. 2001;13:265–73.
4. Tremaine W, Dixon P. A long-term study of 277 cases of equine sinonasal disease. Part 1: details of horses, historical, clinical and ancillary diagnostic findings. Equine Vet J. 2001;33:274–82.
5. Van Cauwenberge P, Vander Mijnsbrugge A-M, Ingels KJ. The microbiology of acute and chronic sinusits and otitis media: a review. Eur Arch Otorhinolaryngol. 1993;250:S3–6.
6. Brook I. Microbiology of acute and chronic maxillary sinusitis associated with an odontogenic origin. Laryngoscope. 2005;115:823–5.
7. Kennedy R, Lappin DF, Dixon PM, Buijs MJ, Zaura E, Crielaard W, et al. The microbiome associated with equine periodontitis and oral health. Vet Res. 2016;47:49.
8. Bienert A, Bartmann C, Verspohl J, Deegen E. Bakteriologische Befunde bei endodontischen und apikalen Erkrankungen der Backenzähne des Pferdes. Dtsch Tierarztl Wochenschr. 2003;110:358–61.
9. Bartmann C, Peters M, Amtsberg G, Deegen E. Dentogene Sinusitis durch Gramnegative Anaerobier beim Pferd. Tierärztl Prax Großtiere. 2002;30:178–83.
10. Gergeleit H, Bienert-Zeit A, Ohnesorge B. Cytological and microbiological examination of secretions from the paranasal sinuses in horses and other species. J Equine Vet Sci. 2018;61:22–31.
11. Igarashi Y, Tada T, Shimura J, Ukita T, Inoue H, Maetani I, Sakai Y. A new cannula with a flexible tip (Swing Tip) may improve the success rate of endoscopic retrograde cholangiopancreatography. Endoscopy. 2002;34:628–31.
12. Woodford N, Lane J. Long-term retrospective study of 52 horses with sinunasal cysts. Equine Vet J. 2006;38:198–202.
13. Schumacher J, Honnas C, Smith B. Paranasal Sinusitis complicated by inspissated exudate in the ventral conchal sinus. Vet Surg. 1987;16:373–7.
14. Mason B. Empyema of the equine paranasal sinuses. J Am Vet Med Assoc. 1975;167:727–31.
15. Cohen NA. Sinonasal mucociliary clearance in health and disease. Ann Otol Rhinol Laryngo. 2006;115:20–6.
16. Carvallo FR, Uzal FA, Diab SS, Hill AE, Arthur RM. Retrospective study of fatal pneumonia in racehorses. J Vet Diagn Invest. 2017;29:450–6.
17. Arroyo M, Slovis N, Moore G, Taylor S. Factors associated with survival in 97 horses with septic pleuropneumonia. J Vet Intern Med. 2017;31:894–900.
18. Timoney JF. The pathogenic equine streptococci. Vet Res. 2004;35:397–409.
19. Waller A. Science-in-brief: *Streptococcus zooepidemicus*: a versatile opportunistic pathogen that hedges its bets in horses. Equine Vet J. 2017;49:146–8.
20. Stoughton WB, Gold J. *Streptococcus equi* subsp. *zooepidemicus* pleuropneumonia and peritonitis in a dromedary camel (Camelus dromedarius) calf in North America. J Am Vet Med Assoc. 2015;247:300–3.
21. Rasmussen CD, Haugaard MM, Petersen MR, Nielsen JM, Pedersen HG, Bojesen AM. *Streptococcus equi* subsp. *zooepidemicus* isolates from equine infectious endometritis belong to a distinct genetic group. Vet Res. 2013;44:26.
22. Decaro N, Mari V, Larocca V, Losurdo M, Lanave G, Lucente MS, et al. Molecular surveillance of traditional and emerging pathogens associated with canine infectious respiratory disease. Vet Microbiol. 2016;192:21–5.
23. Pelkonen S, Lindahl SB, Suomala P, Karhukorpi J, Vuorinen S, Koivula I, et al. Transmission of *Streptococcus equi* subspecies *zooepidemicus* infection from horses to humans. Emerg Infect Dis. 2013;19:1041.
24. Erol E, Locke SJ, Donahoe JK, Mackin MA, Carter CN. Beta-hemolytic *Streptococcus* spp. from horses: a retrospective study (2000–2010). J Vet Diagn Invest. 2012;24:142–7.
25. Mantovani K, Bisanha AA, Demarco RC, Tamashiro E, Martinez R, Anselmo-Lima WT. Maxillary sinuses microbiology from patients with chronic rhinosinusitis. Braz J Otorhinolaryngol. 2010;76:548–51.
26. Kern I, Bartmann C, Verspohl J, Rohde J, Bienert-Zeit A. Bacteraemia before, during and after tooth extraction in horses in the absence of antimicrobial administration. Equine Vet J. 2016;49:178–82.
27. Cehak A, Von Borstel M, Gehlen H, Feige K, Ohnesorge B. Necrosis of the nasal conchae in 12 horses. Vet Rec. 2008;163:300–2.
28. Wilson MT, Hamilos DL. The nasal and sinus microbiome in health and disease. Curr Allergy Asthma Rep. 2014;14:485.
29. Boase S, Foreman A, Cleland E, Tan L, Melton-Kreft R, et al. The microbiome of chronic rhinosinusitis: culture, molecular diagnostics and biofilm detection. BMC Infect Dis. 2013;13:210.
30. Feazel LM, Robertson CE, Ramakrishnan VR, Frank DN. Microbiome complexity and *Staphylococcus aureus* in chronic rhinosinusitis. Laryngoscope. 2012;122:467–72.
31. Cryer J, Schipor I, Perloff JR, Palmer JN. Evidence of bacterial biofilms in human chronic sinusitis. ORL. 2004;66:155–8.

3.2 Publikation III

Cytological examination of secretions from the paranasal sinuses in horses

Published in: Journal of Equine Veterinary Science
Volume 78, July 2019, Pages 60–68

Received: 12 January 2019
Received in revised form: 9 March 2019
Accepted: 10 March 2019

H. Gergeleit, A. Bienert-Zeit, A. Seemann-Jensen, J. Delarocque, B. Ohnesorge
Clinic for Horses, University of Veterinary Medicine Hannover, Foundation, Hannover, Germany

Abstract

The objective of this study was to determine whether the assessment of cytological features of secretions from the paranasal sinuses represents a useful diagnostic tool in equine sinusitis to distinguish between different etiologies. Secretion samples from 50 horses with sinusitis and 10 healthy horses were taken trans-endoscopically from the drainage angle of the nasomaxillary aperture using a Swing Tip catheter. An additional direct sample from the caudal maxillary sinus was taken from all healthy horses after trephination. A direct sample was obtained from the affected sinus in 19 diseased horses after osteotomy. Samples were smeared on microscope slides and dyed using 'Pappenheim stain' for cytological evaluation.

Smears from horses with sinusitis (primary n = 14, dental n = 28, cyst n = 2, progressive ethmoidal hematoma (PEH) n = 2, traumatic n = 2, malignant neoplasia n = 2) were dominated by moderate to high numbers of neutrophilic granulocytes and moderate numbers of different epithelial cells, whereas samples from healthy sinuses revealed almost exclusively epithelial cells. Smears in dental related sinusitis tended to contain more lytic cellular material than in other conditions. Horses with dental sinusitis also showed a higher intra- and extracellular bacterial burden than horses with other underlying etiologies and healthy horses. The absence of bacteria in light microscopy did not exclude the evidence of pathogenic bacteria from bacterial culture in every case. Case numbers were too small to evaluate the diagnostic value for neoplastic conditions or PEH. In conclusion, cytological examination of secretions from the paranasal sinuses can provide an additional but limited ancillary diagnostic value in equine sinusitis.

Journal of Equine Veterinary Science 78 (2019) 60–68

Contents lists available at ScienceDirect

Journal of Equine Veterinary Science

journal homepage: www.j-evs.com

Original Research

Cytological Examination of Secretions From the Paranasal Sinuses in Horses

Hauke Gergeleit*, Astrid Bienert-Zeit, Anja Seemann-Jensen, Julien Delarocque, Bernhard Ohnesorge

Clinic for Horses, University of Veterinary Medicine Hannover, Foundation, Hannover, Germany

ARTICLE INFO

Article history:
Received 12 January 2019
Received in revised form
9 March 2019
Accepted 10 March 2019
Available online 21 March 2019

Keywords:
Cytology
Endoscopy
Equine
Nasomaxillary aperture
Sinusitis

ABSTRACT

The objective of this study was to determine whether the assessment of cytological features of secretions from the paranasal sinuses represents a useful diagnostic tool in equine sinusitis to distinguish between different etiologies. Secretion samples from 50 horses with sinusitis and 10 healthy horses were taken transendoscopically from the drainage angle of the nasomaxillary aperture using a Swing Tip catheter. An additional direct sample from the caudal maxillary sinus was taken from all healthy horses after trephination. A direct sample was obtained from the affected sinus in 19 diseased horses after osteotomy. Samples were smeared on microscope slides and dyed using "Pappenheim stain" for cytological evaluation. Smears from horses with sinusitis (primary n = 14, dental n = 28, cyst n = 2, progressive ethmoidal hematoma [PEH] n = 2, traumatic n = 2, malignant neoplasia n = 2) were dominated by moderate to high numbers of neutrophilic granulocytes and moderate numbers of different epithelial cells, whereas samples from healthy sinuses revealed almost exclusively epithelial cells. Smears in dental-related sinusitis tended to contain more lytic cellular material than in other conditions. Horses with dental sinusitis also showed a higher intracellular and extracellular bacterial burden than horses with other underlying etiologies and healthy horses. The absence of bacteria in light microscopy did not exclude the evidence of pathogenic bacteria from bacterial culture in every case. Case numbers were too small to evaluate the diagnostic value for neoplastic conditions or PEH. In conclusion, cytological examination of secretions from the paranasal sinuses can provide an additional but limited ancillary diagnostic value in equine sinusitis.

1. Introduction

Paranasal sinusitis is a frequently described disease in horses that often has a chronic progression [1]. The preliminary diagnostic intentions in human and equine medicine are distinct as dental-related, and other secondary sinusitis play a minor role in human medicine with most of the cases being classified as acute or chronic primary sinusitis [2]. The diagnosis of primary sinusitis in equine medicine can only be made if all possible secondary etiologies are ruled out, and therefore, the initial focus lies herein when a suspected horse is presented [3]. However, if cases resolve after symptomatic treatment, they are usually presumed to have had primary sinusitis. Many cases of equine sinusitis are secondary to other disorders, and the underlying cause can often only be found by a combination of several imaging diagnostic tools, including endoscopy, radiography, and sinuscopy [1,4]. Cases that remain uncertain may require computed tomography (CT) to confirm the diagnosis, as it provides a detailed visualization of structures without superimposition [5,6]. Scintigraphy and magnetic resonance imaging can be used but these imaging modalities are less advantageous when compared to CT [6,7].

It is concluded that cytology has not been proven to be an adequate tool in the diagnosis of rhinosinusitis in humans, but a biopsy might be indicated to exclude neoplastic or severe

Animal welfare/ ethical statement: The experimental protocol for the control group was approved by the Lower Saxony State Office for Consumer Protection and Food Safety (reference no. 33.19-42502-04-16/2212). Horse owners gave informed consent for their animals' inclusion in the study.

Conflict of interest statement: The authors report no financial or other conflicts related to this work.

* Corresponding author at: Clinic for Horses, University of Veterinary Medicine Hannover, Foundation, Hannover 30559, Germany.

E-mail address: hauke.gergeleit@tiho-hannover.de (H. Gergeleit).

https://doi.org/10.1016/j.jevs.2019.03.003
0737-0806/© 2019 Elsevier Inc. All rights reserved.

conditions, such as vasculitides [8]. Former studies have used mainly leucocytes and especially neutrophil counts in samples that have been harvested from the middle nasal meatus to distinguish between healthy and diseased sinuses [9,10]. These cells have also been used to distinguish a pathogenic from a saprophytic organism isolated from secretions [11,12]. The cytological features of secretions from the paranasal sinuses have rarely been assessed in equine medicine and only a few authors suggest sample collection for cytological examination [4,13], but they do not comment on their interpretation. Bertone et al. state that cytologic examination of secretions aspirated from the paranasal sinuses can be helpful to verify the presence of tumor cell exfoliation but mention that results must be evaluated carefully [14].

However, cytological examination of secretions from the lower airways of horses is a standard clinical procedure to assess character, severity, and progression or remission of pulmonary diseases and can, therefore, help to confirm a suspected diagnosis and decide on the most promising therapy. An examination of bronchoalveolar lavage fluid represents a sensitive method to diagnose diffuse diseases, such as chronic obstructive bronchitis and interstitial pneumopathies. Tracheobronchial secretion analysis is useful for generalized diseases as it provides insights into the conditions of the respiratory epithelium and the effectiveness of mucociliary clearance [15–18].

Furthermore, cytological examination of body fluids and cellular secretions displays a useful ancillary diagnostic tool in equine medicine in several conditions including pleural, abdominal, synovial, and cerebral fluids [19]. Effusions from body cavities can especially help to distinguish between inflammatory, infectious, and neoplastic conditions, and therefore, their interpretation can determine the prognosis [20–22]. The making of cytological slides requires only sparse equipment, and the stained smears are usually available within a short time, which allows direct evaluation and the results can also contribute directly to the surgical decision especially if neoplastic conditions are evident, although they are relatively uncommon.

Under these circumstances, the question whether cytological examination of paranasal sinus secretions might also be helpful in making an etiological diagnosis is raised. Therefore, the preliminary aim of this study was to describe the cytological findings of secretions from the paranasal sinuses in healthy horses and those affected by sinusitis. The second aim was to assess whether these findings allow one to draw conclusions regarding the underlying etiology of sinusitis and whether this approach may display a feasible ancillary diagnostic tool in equine paranasal sinus diseases.

2. Material and Methods

2.1. Patient Group

The patient group consisted of 50 horses, including 37 Warmbloods, 7 ponies, 3 Icelandic horses, 1 English Thoroughbred, 1 Arabian Thoroughbred, and 1 Coldblood (26 mares, 22 geldings, and 2 stallions) with an age (mean $\pm$ s.d.) of 12.7 $\pm$ 6.4 years (range 1–27 years), presented to the Clinic for Horses of the University of Veterinary Medicine Hannover, Foundation, between October 2016 and September 2017 with a history of unilateral nasal discharge. They were diagnosed with a paranasal sinus disease. The history of all horses included the character and duration of clinical signs as well as previous medical and surgical treatments. All horses underwent a detailed clinical examination and a specific examination of the head [23]. This included assessment of presence, odor, and character of nasal discharge, airflow from the nares, symmetry of the head, and palpation of the mandibular lymph nodes and percussion of the paranasal sinuses to detect dullness.

2.2. Further Diagnostics and Sampling

Subsequent transnasal endoscopic examination of the respiratory tract was performed using an Olympus video endoscope (PCF-H180AL; Olympus Deutschland GmbH, Hamburg, Germany). All horses were restrained in stocks and sedated using detomidine (Cepesedan; CP-Pharma Handelsgesellschaft mbH, Burgdorf, Germany, 0.01 mg/kg bwt i.v.) combined with butorphanol (Butorgesic; CP-Pharma Handelsgesellschaft mbH, Burgdorf, Germany, 0.01 mg/kg bwt i.v.). Particular attention was paid to the nasal conchae, the nasal meatus, with special regard to the caudal aspect of the middle nasal meatus (i.d. "drainage angle"), and the ethmoid region. Presence, quantity, and nature of any secretion draining off the nasomaxillary aperture along the common sinonasal canal were noted. The endoscope was positioned in the middle nasal meatus rostral to the secretion from the paranasal sinuses. If swelling restricted placement of the endoscope in the middle nasal meatus, it was positioned in the common nasal meatus slightly rostrally and ventrally to the distal end of the middle nasal meatus. A flexible, wire-guided catheter (Olympus Swing Tip$_{TM}$ PR-233Q; Olympus Deutschland GmbH, Hamburg, Germany [24]) was advanced through the working channel of the endoscope. Its distal tip was angulated using the handle attached until the catheter was positioned in the drainage angle and advanced towards the common sinonasal canal for two to three centimeters. At that location, secretion was aspirated into the catheter [25]. The catheter was removed before the endoscopic examination was completed to exclude other diseases.

Depending on the history, clinical signs and findings during the endoscopic examination, further diagnostic tools, including examination of the oral cavity, radiography, and CT of the head, were combined to make an etiological diagnosis.

After the diagnostics and transendoscopic sampling, 19 diseased horses that required intensive sinus irrigation or mass removal underwent trephination or bone flap surgery, respectively. On these occasions, a sterile dry cotton swab was inserted through the opening of the osteotomy and a sample of the secretion was taken directly from the sinus affected.

2.3. Control Group

The control group consisted of 10 horses, including 6 Warmbloods, 3 Trotters, and 1 Icelandic horse (9 mares and 1 gelding) with an age (mean $\pm$ s.d.) of 16.6 $\pm$ 7.3 years (range 2–30 years) owned by the Clinic for Horses of the University of Veterinary Medicine Hannover, Foundation. The horses had no known history of sinonasal diseases. They each underwent a detailed clinical and radiographic examination of the head. The horses were put under general anesthetic for an unrelated terminal study. All procedures were approved by the Lower Saxony State Office for Consumer Protection and Food Safety (reference no. 33.19-42502-04-16/2212). Signs of sinonasal disorders were excluded during the endoscopic examination before secretion samples were aspirated using the technique described previously. The left side was used for transendoscopic sampling in all control horses. Afterwards, the ipsilateral frontal sinus was trephined, and a mucosal smear was collected from the caudal maxillary sinus with a sterile dry cotton swab and a cytobrush.

An additional sample was taken transendoscopically from the contralateral, clinically inconspicuous side of 35 horses from the patient group. Intraindividual findings between the affected and the nonaffected side were compared.

Table 1
Scoring system used to evaluate the cytological findings semiquantitatively.

Definition	Score
No cellular content	0
Few cells in whole survey path	1
Several cells in whole survey path	2
Few cells on every straight line	3
Several cells on every straight line	4
Cell type dominating the survey path	5

2.4. Cytological Processing and Examination

The secretion samples were smeared onto microscope slides for cytological examination. Therefore, the samples taken transendoscopically were blown out of the catheter with a syringe onto one microscopic slide and streaked out against a second microscopic slide. The cotton swab or cytobrush samples were streaked out onto microscopic slides in a rotating manner. All smears were air-dried and dyed using a combination of the May-Grünwald and Giemsa stains (i.e., "Pappenheim stain"). The smears were examined in a meandering pattern under a light microscope (Zeiss Axio Scope.A1; Carl Zeiss Microscopy GmbH, Jena, Germany) with 40x and 500x magnification. At least 10 straight lines were considered. Cytological findings were evaluated semiquantitatively using the scoring system displayed in Table 1. Intermediate score levels were chosen if appropriate. Microscopic examination was performed by two different observers (H.G. and A.S.J.) twice independently to assess interobserver and intraobserver agreements.

2.5. Statistical Analysis

Statistical analysis was performed with R 3.5.1 [26]. Kruskal–Wallis tests were used to assess the differences of score values for selected parameters (Table 2) of the cytological examination between healthy sinuses, and primary and dental sinusitis. The *P*-values were adjusted for multiple testing using the Holm method. Dunn's test with Bonferroni correction was used as a post hoc test for pairwise comparisons across the different sinus conditions.

A multinomial model predicting the etiology and including the variables identified as significant by the Kruskal-Wallis tests was fitted and tuned using the R packages nnet [27] and caret [28]. The model accuracy was determined using a 7-fold repeated cross-validation with 10 repeats. This means that 11 of the 77 samples were excluded when building the model and then used to test the model. This was repeated until all samples had been used 10 times in random groups of 7 to test the model.

Fisher's exact test was used to assess the occurrence of lytic cells in dependence on the presence of bacteria. Cohen's weighted kappa was used to measure interobserver and intraobserver agreements in cytological evaluations. Weights were set as the squared distance from perfect agreement. Agreements were classified according to suggestions by McHugh [29]. *P*-values < 0.05 were considered statistically significant for all calculations. Smears from horses with sinusitis that had an underlying etiology other than primary or dental sinusitis were evaluated descriptively because of small sample numbers.

3. Results

The etiological diagnosis for the sinusitis was a primary sinusitis in 14 horses; a dental sinusitis in 28 horses; and malignant neoplasia, progressive ethmoidal hematoma (PEH), paranasal sinus cyst, and traumatic sinusitis in two horses each.

3.1. Transendoscopic Sampling

Neutrophilic granulocytes were the predominant cell type in all diseased horses with higher scores in primary sinusitis (median of

Table 2
Cytological findings of transendoscopic samples and distribution of scores for selected parameters among healthy sinuses, primary sinusitis, and dental sinusitis.

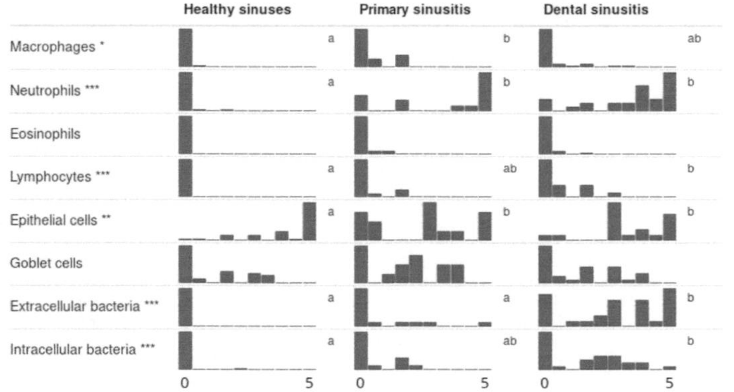

Kruskal-Wallis tests were used to determine whether the cell populations mentioned were present in similar patterns across the different conditions.Statistical significance is indicated as *$P < .05$, **$P < .01$, and ***$P < .001$, and *P*-values are corrected for multiple testing using the Holm method. When the results were significant, a Dunn test for pairwise comparisons with Bonferroni correction was conducted. Different letters alongside the histograms of one row indicate a significant result of the pairwise comparison ($P < .05$). As an example, macrophages differ significantly in their prevalence in smears from horses with primary sinusitis compared to smears from healthy sinuses. No significant differences in macrophages are found in smears from dental sinusitis compared to either healthy sinuses or primary sinusitis.

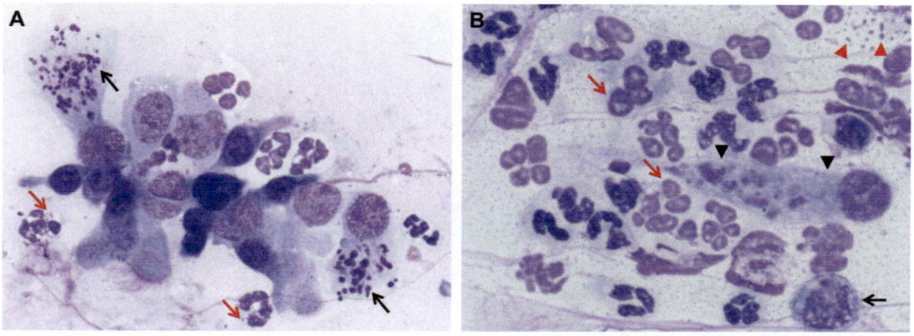

Fig. 1. Primary sinusitis, (A): squamous and ciliated epithelial cells, goblet cells with cytoplasmic granules (black arrows), neutrophils with intracellular bacteria (red arrows) and low numbers of extracellular bacteria; (B): neutrophilic granulocytes, some with a degenerative character (red arrows), goblet cell (black arrowheads), macrophage (black arrow), stain precipitate (red arrowheads) but no visible bacteria (Pappenheim's stain; original magnification 500X).(For interpretation of the references to color in this figure legend, the reader is referred to the Web version of this article.)

5, Fig. 1) than in dental sinusitis (median of 4, Fig. 2). Degenerative neutrophils were frequently observed in both groups (Figs. 1B and 2A), with these cells tending to have a more lytic character in cases of dental sinusitis than in primary sinusitis, but this observation could not be confirmed statistically. However, lytic cells were significantly more frequently observed if bacteria were present in the microscopic samples ($P < .05$).

Macrophages were observed less frequently and in smaller numbers in diseased sinuses with 2 being the highest score given by both observers (Fig. 2A).

Other inflammatory cells, such as lymphocytes and eosinophilic granulocytes, were only sporadically identified in small numbers (Fig. 2B), with the latter showing the highest occurrence in a horse with a PEH (Fig. 3B).

Epithelial cells were frequently found in most of the smears and included squamous epithelial cells, ciliated, mostly columnar, epithelial cells, and goblet cells in varying amounts (Fig. 1A). The median score for epithelial cells was 3 in samples from diseased sinuses.

Overall, there were no statistically significant differences regarding the scoring for inflammatory cells between primary and dental sinusitis in any of the evaluations.

However, bacteria were significantly more frequently present in smears from dental sinusitis than in other etiologies ($P < .01$). Intracellular bacteria were scored higher in dental than in primary sinusitis, although these values were not significant ($P = .088$, Fig. 2 and Table 2). Neutrophils especially showed high bacterial load in dental sinusitis (Fig. 2A).

The distribution of the scores for different cytological findings among healthy sinuses, and primary and secondary dental sinusitis are summarized in Table 2.

Smears from horses with an underlying etiology other than primary or dental were also dominated by moderate to high numbers of neutrophils (median of 4, Figs. 3, 4 and 7) and numerous epithelial cells (median score of 3). The secretion samples from both horses with PEH and traumatic sinusitis had moderate to high scores for erythrocytes (Figs. 3 and 4). In PEH, light microscopy revealed large numbers of inflammatory cells in the PEH including macrophages, neutrophils, and a higher number of eosinophilic granulocytes than in other etiologies (Fig. 3). Intracellular pigments were found in only one horse with a PEH (Fig. 3A). Tissue samples for histopathological evaluations revealed a squamous cell carcinoma in one horse and a malignant blastoma in another horse. The cytological smear from the horse with the

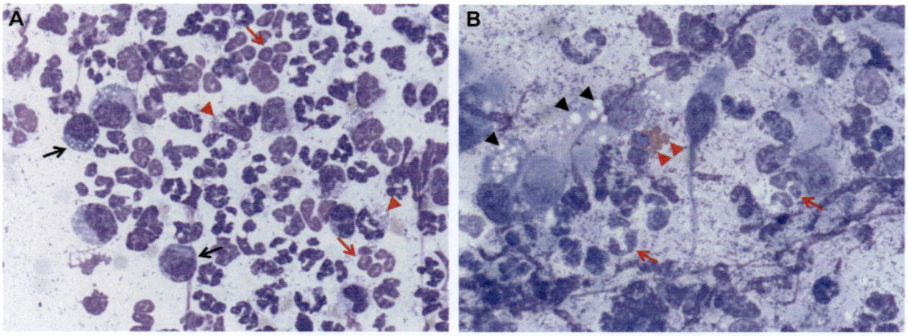

Fig. 2. Dental sinusitis, (A): neutrophilic granulocytes with high numbers of intracellular bacteria (red arrowheads), degenerative neutrophils (red arrows), macrophage (black arrow); (B): neutrophilic granulocytes (red arrows), eosinophilic granulocyte (double red arrowheads), and vacuolated epithelial cells (black arrowheads) surrounded by high numbers of extracellular bacteria (Pappenheim's stain; original magnification 500X).(For interpretation of the references to color in this figure legend, the reader is referred to the Web version of this article.)

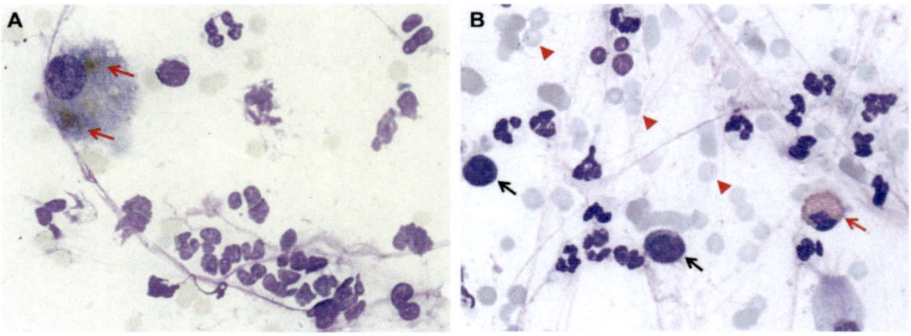

Fig. 3. Progressive ethmoidal hematoma, (A): macrophage with intracytoplasmic pigments, probably hemosiderin (red arrows); (B): macrophages (black arrows), partly clotted erythrocytes (red arrowheads), neutrophils and an eosinophilic granulocyte (red arrow). Eosinophils were a frequent occurrence in this patient (Pappenheim's stain; original magnification 500X).(For interpretation of the references to color in this figure legend, the reader is referred to the Web version of this article.)

squamous cell carcinoma was dominated by neutrophils, of which many had a lytic character and large numbers of extracellular and intracellular bacteria. Neoplastic changes could not be verified in the few epithelial cells that were also present. The smear from the malignant blastoma showed mainly degenerative neutrophils and epithelial cells (Fig. 7B).

Smears taken from the healthy control site in 35 horses were dominated by squamous epithelial cells, ciliated epithelial cells, and goblet cells (Fig. 5). Neutrophilic granulocytes occurred almost exclusively in smears from diseased horses, whereas only a very few were found in four smears from healthy controls ($P < .001$, Fig. 5A). Macrophages were found comparably more frequently than other inflammatory cells in healthy controls, but scores were still lower than in the horses with a dental sinusitis ($P = .0501$) and significantly lower than in those with a primary sinusitis ($P < .05$).

The intrarater agreement for repeated measurements was strong and better for neutrophils, epithelial cells, bacteria, and intracellular bacteria in both observers (Fig. 8). It was moderate for erythrocytes and eosinophils in one observer and weak to minimal in macrophages and lymphocytes in both observers, respectively.

The inter-rater agreement was strong for the evaluation of neutrophils and moderate for bacteria. For the other evaluations, it was weak and less.

3.2. Direct Sampling

Trephination was performed in four horses with a primary sinusitis, 13 horses with a dental sinusitis, and one horse with a sinus cyst, whereas one horse with a PEH underwent bone flap surgery.

Inspissated secretions with a caseous character, as has been seen frequently in horses with a chronic dental sinusitis, did not lead to identifiable cells in smears in most of the cases (Fig. 6A). The material consisted mainly of lytic or necrotic debris and high to massive numbers of bacteria. Cellular lysis and necrosis showed less progression in samples that were less inspissated as in some cases of dental and primary sinusitis (Fig. 6B).

Epithelial cells were found less frequently in trephination samples (Scores ≤ 3) compared to the endoscopic samples from diseased horses (Scores ≤ 5).

Ten healthy horses underwent endoscopic examination and ipsilateral sinus trephination. None of the trephined sinuses contained accumulations of secretions, and the mucosa was pale pink with a slightly moist surface. Endoscopic and direct samples were almost exclusively dominated by epithelial cells (median score of 5). Cytobrush smears revealed more cellular material than cotton swab samples but brush samples resulted in more cellular detritus.

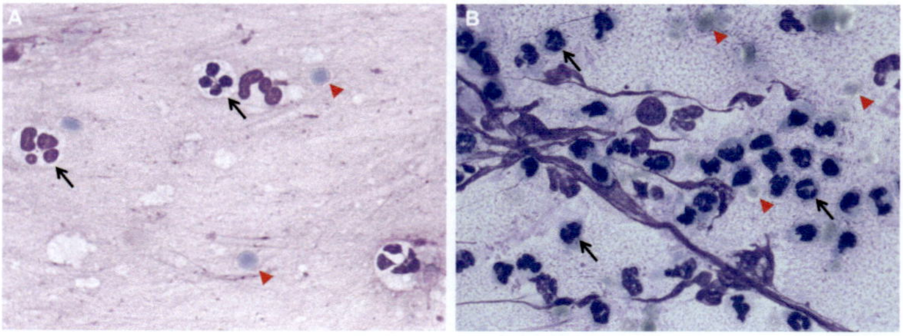

Fig. 4. Traumatic sinusitis, (A): segmented neutrophilic granulocytes (black arrows) and some erythrocytes (red arrowheads) surrounded by a heterogeneous mat of pink-staining mucin; (B): higher density of neutrophils (black arrows) in another horse, erythrocytes (red arrowheads), and mucous strings (Pappenheim's stain; original magnification 500X).(For interpretation of the references to color in this figure legend, the reader is referred to the Web version of this article.)

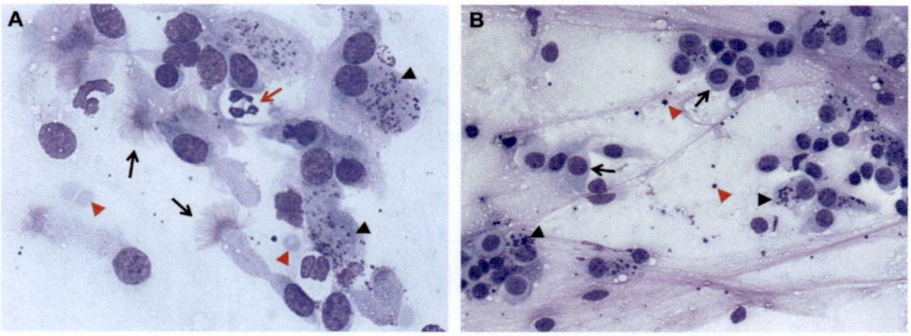

Fig. 5. Smear from the "drainage angle" of a healthy control side, (A): secretory goblet cells with cytoplasmic mucin granules (black arrowheads), ciliated columnar epithelial cells (black arrows), single neutrophilic granulocyte (red arrow), and few erythrocytes (red arrowheads); (B): goblet cells (black arrowheads), squamous epithelial cells (black arrows), and free mucin granules (red arrowheads) (Pappenheim's stain; original magnification 500X).(For interpretation of the references to color in this figure legend, the reader is referred to the Web version of this article.)

Only a few neutrophils and macrophages were found in two of these horses in the whole survey path (maximum scores of 1). Direct samples were slightly contaminated by blood from the skin incision in three horses.

3.3. *Predicting the Etiology*

The average accuracy of the multinomial model was 70.8%. The confusion matrix of the model is presented in Table 3. These results indicate that while the model has a high accuracy in identifying healthy samples, it has only a moderate performance in detecting a dental sinusitis and poor performance for primary sinusitis. The latter is equally often identified as healthy and primary but even more often as a dental sinusitis.

4. Discussion

Transnasal endoscopy is essential to examine a horse thoroughly that is presented with nasal discharge to locate the origin of the secretions [4].

Transendoscopic sampling of secretions from the paranasal sinuses has only once been reported in equine medicine [24]. This is the first study that has used a wire-guided, flexible catheter to facilitate directed sampling. This approach allows the diagnostic imaging and sample collection for further evaluation in one step. Other procedures that have been described previously to gain samples include fine needle aspiration through the overlying bone and sinus trephination [4]. The surgical and cosmetic outcome of these procedures are most commonly satisfying but opening the paranasal sinuses bears the risk of wound infection and delayed secondary healing [30].

There are no studies to date that have focused on the cytological examination of secretions from healthy and diseased equine paranasal sinuses, and only a few authors suggest sampling for such purposes without giving further information on their interpretation for the practitioner [4,13]. Some of the few microscopic photographs of stained exudates from paranasal sinuses are published by Cowell and Tyler [19], but the underlying etiology of the samples illustrated is not mentioned. This is the first attempt to describe, compare, and depict the cytological features of paranasal sinus secretions in healthy horses and in different etiologies for sinusitis.

Healthy paranasal sinuses are covered by respiratory pseudostratified, columnar, ciliated, or cuboidal transitional epithelium that contains varying numbers of goblet cells [19,31]. The mucosa is under constant and dynamic modulation to maintain mucociliary clearance in both, health and disease, and therefore these cells were found in samples from healthy and diseased sinuses. The result of an increase in goblet cells and increased amounts of mucin

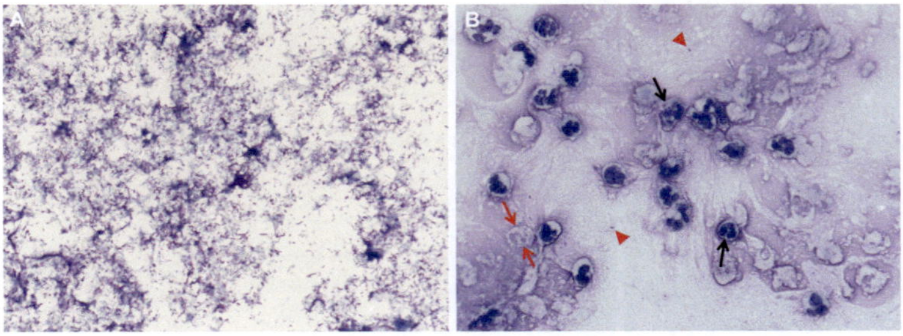

Fig. 6. Direct samples after trephination, (A): dental sinusitis—necrotic debris and bacteria but no identifiable cells; (B): primary sinusitis—degenerate neutrophils (black arrow), debris from lysed neutrophils (red arrows), and small numbers of extracellular coccoid bacteria (red arrowheads) (Pappenheim's stain; original magnification 500X).(For interpretation of the references to color in this figure legend, the reader is referred to the Web version of this article.)

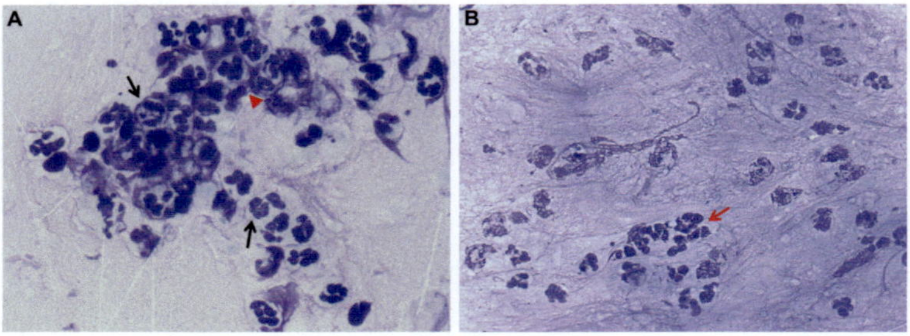

Fig. 7. (A): Paranasal sinus cyst—domination of neutrophilic granulocytes (black arrows) with few bacteria (red arrowhead); (B): malignant blastoma (diagnosed by histopathology)—lytic neutrophils (red arrow) and debris from lysed cells but no intact cells (Pappenheim's stain; original magnification 500X).(For interpretation of the references to color in this figure legend, the reader is referred to the Web version of this article.)

produced because of mucosal irritation, such as infectious organisms or other noxious stimuli, can be seen cytologically as mats of homogenous, pink to red-staining material or finely mottled pink background (Fig. 4A). Mucosal inflammation with decreased efficiency of mucociliary clearance and increased amounts of secretions and inflammatory cell extravasation [32,33] led to samples that varied in their quantity and viscosity. This causes a greater risk of contamination of the sample with cells from the nasal mucosa during advancement of the endoscope and the catheter, especially when secretions were of a lower quantity, displaying a limitation of the study. Only a few drops of serous secretions could be aspirated from the mucosal surface in healthy control horses using the transendoscopic approach. The nasal mucosa, including the mucosa in the area of the nasomaxillary aperture, has been proven to contain higher numbers of goblet cells than that within the paranasal sinuses [31,34]. Therefore, the interpretation, especially of epithelial cell contents, must be made with caution. To put this risk into perspective, samples collected transendoscopically and directly were compared with the latter presenting smaller amounts of epithelial cells, suggesting contamination of the endoscopic samples due to scratching on the mucosal surface of the nasomaxillary aperture during advancement of the catheter. Hence, exfoliation of epithelium may play a minor role in the cellular composition of secretions in sinusitis.

The features of the inflamed mucosa were similar between dental sinusitis and chronic primary sinusitis in a histological study on horses affected by sinusitis and did not allow one to draw conclusions regarding either of the underlying etiologies [31]. These findings are in accordance with those of the present study, as we detected similar degrees of neutrophils, macrophages, and desquamated epithelial cells in both groups of diseased horses but only very few inflammatory cells in healthy controls. Nevertheless, an obvious difference between both diseased groups was the abundance of intracellular and extracellular bacterial contents with significantly higher scores in dental sinusitis. This indicates that a microscopic examination of a stained smear can help to diagnose the underlying etiology within a short time and before the microbiological results are available. Microbiological cultivation is essential to identify the organisms involved, as identification of particular bacterial pathogens based on their cytological appearance is not reliable [19]. In another part of this study, the secretion samples were examined microbiologically [25]. To some extent, the

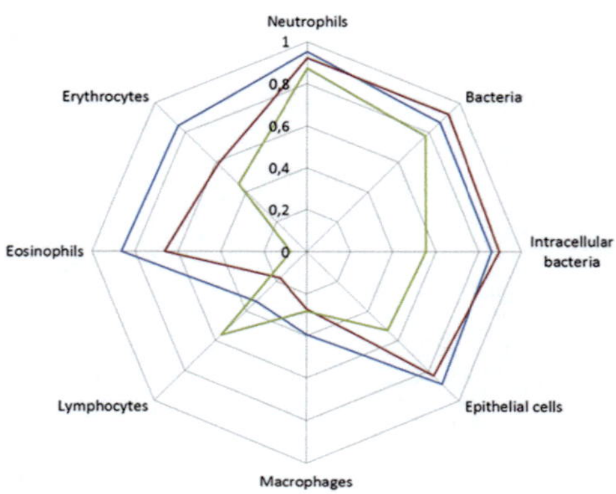

	Intra-rater I	Intra-rater II	Inter-rater
Neutrophils	0,9521	0,9221	0,8721
Bacteria	0,8721	0,9293	0,7764
Intracellular bacteria	0,8602	0,8971	0,5545
Epithelial cells	0,8907	0,8324	0,5292
Macrophages	0,3941	0,2718	0,2822
Lymphocytes	0,3352	0,1755	0,5638
Eosinophils	0,8627	0,6614	0,0765
Erythrocytes	0,8478	0,5862	0,4509

Fig. 8. Inter- and intra-rater agreements for cellular evaluations: Agreements are indicated in decimal numbers.

H. Gergeleit et al. / Journal of Equine Veterinary Science 78 (2019) 60–68

Table 3
Confusion matrix of the multinomial model.

Prediction	Reference		
	Healthy sinuses	Primary sinusitis	Dental sinusitis
Healthy sinuses	42.5% (93.4%)	5.3% (29.1%)	3.2% (8.8%)
Primary sinusitis	2.6% (5.7%)	5.3% (29.1%)	10.1% (27.8%)
Dental sinusitis	0.4% (0.9%)	7.5% (41.2%)	23.0% (63.4%)

The values represent the percentual average cell counts across the 7-fold cross-validation with 10 repetitions. The percentage of counts relative to the reference is presented in brackets. As an example, 93.4% of the healthy samples were classified as such. By contrast, only 29.1% of the samples from primary sinusitis could be detected as such.

cytological results show a good correlation with the microbiological results. Samples from horses with dental sinusitis showed a higher median number of isolated bacterial species, a higher diversity and bacteria were more frequently isolated in high numbers than in primary sinusitis [25]. These findings are reflected in the higher abundance of bacteria in the cytologic examination of the smears from horses with dental sinusitis.

Furthermore, all samples from diseased horses revealed growth of at least one bacterial species, with most of the samples showing mixed bacterial growth [25]. Notwithstanding this, not all smears from horses with dental sinusitis and only 9 of 14 horses with primary sinusitis contained bacteria shown using light microscopy. Therefore, we conclude that the absence of bacteria in the cytologic examination does not exclude the presence of pathogenic microorganisms in the paranasal sinuses. The occurrence of lytic cells in our study was associated with the detection of bacteria using light microscopy. Even if bacteria are not observed directly, the presence of degenerate neutrophils in a cytological smear suggests infection with toxin-producing bacteria. These cells appear swollen, lose their nuclear segmentation, and have lighter, pink-staining chromatin [19] (Figs. 1B and 2A). Not only free mucin granules, cilia from denuded epithelial cells, and stain precipitate but also necrotic debris can be mistaken for bacteria. In addition, one should scrutinize the microbiological results critically when microscopy of the same sample does not show bacterial abundance as some bacterial species might show extensive growth under laboratory conditions albeit they play a minor role in the disease. Furthermore, high numbers of neutrophilic granulocytes do not necessarily indicate an infection with pathogenic bacteria, as we identified many neutrophils in horses with traumatic sinusitis, whereas the correlating bacterial culture only revealed small numbers of alpha-hemolytic streptococci and coagulase-negative staphylococci. The pure number of neutrophils does not seem to correlate with the severity of the disease but rather their degenerative and lytic progression under severe bacterial infection. Nevertheless, Araujo et al. used the number of neutrophils to distinguish between pathogenic and saprophytic organisms [12]. This should be evaluated critically as it may lead to false-negative and false-positive results as our findings suggest.

The findings from both horses with PEH correlate with findings in the histopathological study by Tremaine et al. [31]. Similar to previous descriptions, we could identify intracellular pigments that are most probably hemosiderin, in association with sanguineous secretions from a horse with a PEH but it could not be described in traumatic sinusitis [31]. Therefore, the presence of hemosiderin might be an indicating parameter to distinguish between these two etiologies as long endoscopic findings are uncertain. The Prussian blue stain can be used to prove the presence of hemosiderin. It remains unclear whether these findings will be representative for a larger population because of small sample sizes (two cases each).

It was not possible to aspirate secretions directly from the trephined sinuses in healthy horses. Hence, sampling had to be performed as a mucosal smear. We conclude that if secretions exceed a normal moist paranasal sinus mucosa in a horse at rest, it can be regarded as pathologically increased and, therefore, indicates mucosal inflammation. This is supported by the dominating presence of neutrophils in all different etiologies for paranasal sinusitis, whereas healthy controls only revealed very few neutrophils and other inflammatory cells in single cases.

In the assessments of both inter- and intra-rater agreements, better agreements could be achieved for cellular structures that had a higher overall prevalence, such as the occurrence of neutrophils and bacteria in diseased horses and the number of epithelial cells in most of the controls. The weaker agreements for cells that showed a lower overall abundance, such as lymphocytes and eosinophils, might be due to limitations in the methods of this study. As an example, different fields of view have probably been looked at in different evaluations of the same macroscopic slide, which may have resulted in distinction of scoring for specific cells. Moreover, the nature of the secretions is heterogeneous, making the evaluation more prone to false-negative scoring values if cells only appear in small numbers. Although most inflammatory cell types can be identified with a little guidance and practice, the identification of specific cellular features and reactions requires more experience, especially when it comes to differentiating physiological from neoplastic cells.

Although it is interesting to know that the cell populations show significant differences across the types of samples represented here, this does not quantify the diagnostic use of cytological analysis. We could show by using a multinomial model that the main parameters of cytological analysis can help to achieve a rather high overall diagnostic accuracy (above 70%). However, this model is more capable of differentiating healthy from diseased than to identify the exact etiology of the sinusitis. A possible explanation is that neutrophils are the most important predictors of the model. In conclusion, the model performed well when distinguishing healthy from diseased but could not identify the most probable etiology in diseased samples. Consequently, we do not present detailed model explanations here. The findings emphasize that while the cytological examination can be used as an add-on, it is not a substitute for further diagnostic procedures. Nevertheless, these cytological profiles remain relevant to the pathomechanisms of the underlying diseases. Moreover, diseases that are more challenging to diagnose and were not included in the model because of low sample sizes could benefit more from such an approach because they might be accompanied by more characteristic features. A larger study population for each group would be required to evaluate cytological findings securely for underlying etiologies other than odontogenic sinusitis.

5. Conclusions

The study provides evidence of significantly distinct patterns in the cytological appearance of paranasal sinus secretions. Cytological examination can help to detect a sinusitis and to distinguish between certain etiologies. Thereby neutrophil counts and the presence of bacteria are the two main distinctive features that are reliably detectable.

Cytological analysis is not suitable as a single diagnostic instrument, however, may serve as an ancillary tool in equine sinusitis when the findings of oral and radiographic examinations do not lead to a clear etiological diagnosis.

Acknowledgment

The authors acknowledge the contributions of all coworkers at the Clinic for Horses, University of Veterinary Medicine, Hannover, for assistance in sample acquisition.

Authors' contributions: HG contributed to the study design, data collection, study execution, and data analysis and interpretation. ABZ and BO contributed to the study design, study execution, and data analysis and interpretation. ASJ contributed to the study execution and data analysis and interpretation. JD contributed to data analysis and statistical evaluation and interpretation. All authors read and approved the final version of the article.

This research did not receive any specific grant from funding agencies in the public, commercial, or not-for-profit sectors.

References

[1] Waguespack RW, Taintor J. Paranasal sinus disease in horses. Compend Contin Educ Vet 2011;33:E1–12.
[2] Brook I. Sinusitis of odontogenic origin. Otolaryngol Head Neck Surg 2006;135:349–55.
[3] Tremaine W, Dixon P. A long-term study of 277 cases of equine sinonasal disease. Part 1: details of horses, historical, clinical and ancillary diagnostic findings. Equine Vet J 2001;33:274–82.
[4] Beard W, Hardy J. Diagnosis of conditions of the paranasal sinuses in the horse. Equine Vet Educ 2001;13:265–73.
[5] Henninger W, Mairi Frame E, Willmann M, Simhofer H, Malleczek D, Kneissl SM, et al. CT features of alveolitis and sinusitis in horses. Vet Radiol Ultrasound 2003;44:269–76.
[6] Kaminsky J, Bienert-Zeit A, Hellige M, Rohn K, Ohnesorge B. Comparison of image quality and in vivo appearance of the normal equine nasal cavities and paranasal sinuses in computed tomography and high field (3.0 T) magnetic resonance imaging. BMC Vet Res 2016;12:13.
[7] Barakzai S, Tremaine H, Dixon P. Use of scintigraphy for diagnosis of equine paranasal sinus disorders. Vet Surg 2006;35:94–101.
[8] Fokkens WJ, Lund VJ, Mullol J, Bachert C, Alobid I, Baroody F, et al. European position paper on rhinosinusitis and nasal polyps 2012. Rhinol Suppl 2012;23:1–298.
[9] Wilson NW, Jalowayski AA, Hamburger RN. A comparison of nasal cytology with sinus x-rays for the diagnosis of sinusitis. Am J Rhinol 1988;2:55–9.
[10] Gill FF, Neiburger JB. The role of nasal cytology in the diagnosis of chronic sinusitis. Am J Rhinol 1989;3:13–5.
[11] Bhattacharyya N. The cytology and microbiology of persistent paranasal sinus secretions after endoscopic sinus surgery: a controlled study. Laryngoscope 2007;117:2041–4.
[12] Araujo E, Dall C, Cantarelli V, Pereira A, Mariante AR. Microbiology of middle meatus in chronic rhinosinusitis. Braz J Otorhinolaryngol 2007;73:549–55.
[13] Ainsworth D, Cheetham J. Disorders of the respiratory system. In: Reed S, Bayly WM, Sellon DC, editors. Equine internal medicine. 3 ed. St. Louis, MO: Elsevier Health Sciences; 2009.
[14] Bertone J, Biller D, Ruggles A. Diagnostic techniques for evaluation of the paranasal sinuses. Vet Clin North Am Equine Pract 1993;9:75.
[15] Dieckmann M, Deegen E. Klinische bedeutung der Tracheobronchialsekretzytologie. Pferdeheilk 1990;6:101–10.
[16] May A, Gehlen H. Durchführung, Analyse und Aussagekraft von Tracheobronchialsekret (TBS) und Bronchoalveolärer Lavage (BAL) bei Pferden mit Lungenerkrankungen. Pferdeheilk 2009;25:310–20.
[17] Fey K. Der klinische Nutzen zytologischer Untersuchungen von BALF bei der Differenzierung chronischer Bronchitiden des Pferdes. Habil Med Vet Gießen 2004.
[18] Bullone M, Joubert P, Gagné A, Lavoie JP, Hélie P. Bronchoalveolar lavage fluid neutrophilia is associated with the severity of pulmonary lesions during equine asthma exacerbations. Equine Vet J 2018;50:609–15.
[19] Cowell RL, Tyler RD. In: Diagnostic cytology and hematology of the horse. E-Book. St. Louis, MO: Elsevier Health Sciences; 2001.
[20] Mastrorilli C, Cesar F, Joiner K, Wooldridge AA, Christopherson PW. Disseminated lymphoma with large granular lymphocyte morphology diagnosed in a horse via abdominal fluid and transtracheal wash cytology. Vet Clin Pathol 2015;44:437–41.
[21] Duran MC, Starrak G, Dickinson R, Montgomery J. Peritoneal fluid immunocytochemistry used for the diagnosis of a possible case of equine gastrointestinal B-cell lymphoma. Can Vet J 2016;57:601.
[22] Arroyo M, Slovis N, Moore G, Taylor S. Factors associated with survival in 97 horses with septic pleuropneumonia. J Vet Intern Med 2017;31:894–900.
[23] Traub-Dargatz J. Field examination of the equine patient with nasal discharge. Vet Clin North Am Equine Pract 1997;13:561–88.
[24] Igarashi Y, Tada T, Shimura J, Ukita T, Inoue H, Maetani I, et al. A new cannula with a flexible tip (Swing Tip) may improve the success rate of endoscopic retrograde cholangiopancreatography. Endoscopy 2002;34: 628–31.
[25] Gergeleit H, Verspohl J, Rohde J, Rohn K, Ohnesorge B, Bienert-Zeit A. A prospective study on the microbiological examination of secretions from the paranasal sinuses in horses in health and disease. Acta Vet Scand 2018;60:43.
[26] Team RCR. A language and environment for statistical computing. Vienna, Austria: R Foundation for Statistical Computing; 2018. https://www.R-project.org/. [Accessed 3 November 2018].
[27] Venables W, Ripley B. Statistics complements to modern applied statistics with S. 4 ed. New York: Springer; 2002.
[28] Kuhn M. Caret package. J Stat Softw 2008;28:1–26.
[29] McHugh ML. Interrater reliability: the kappa statistic. Biochem Med 2012;22: 276–82.
[30] Barakzai S. Standing sinus surgery in the horse: indications, techniques and complications. Practice 2008;30:252–62.
[31] Tremaine W, Clarke C, Dixon P. Histopathological findings in equine sinonasal disorders. Equine Vet J 1999;31:296–303.
[32] Cohen NA. Sinonasal mucociliary clearance in health and disease. Ann Otol Rhinol Laryngol 2006;115:20–6.
[33] Fey K, Ohnesorge B. Erkrankungen der Nase, Nasennebenhöhlen und Nasenmuscheln. In: Dietz O, Huskamp B, editors. Handbuch pferdepraxis. 4 ed. Stuttgart, Germany: Georg Thieme Verlag; 2016.
[34] Schwieder A. Untersuchung zur histologischen Beschaffenheit der Schleimhaut der Sinus paranasales des Pferdes unter Berücksichtigung von Topographie und Alter. Diss Med Vet Hannover 2018:71–105.

4 Übergreifende Diskussion

In der vorliegenden Studie wurde die mikrobiologische und zytologische Beschaffenheit von Sekreten aus den Nasennebenhöhlen gesunder und erkrankter Pferde beschrieben. Anhand der erhobenen Befunde in Zusammenhang mit der zugrundeliegenden ätiologischen Diagnose der Sinusitiden wurde erstmals der diagnostische Nutzen der Untersuchung dieser Sekrete evaluiert.

4.1 Diskussion von Material und Methode

4.1.1 Probandengut

4.1.1.1 Erkrankte Pferde

In Bezug auf die Krankheitsdauer und medikamentöse Vorbehandlungen war die Gruppe der erkrankten Pferde insgesamt sehr heterogen. Insbesondere vorangegangene antibiotisch wirksame, systemische Behandlungen hatten auf die mikrobiologischen Ergebnisse einen nicht zu kalkulierenden Einfluss. Bei keinem der Pferde wurde vorberichtlich gezielt gegen anaerobe Bakterien behandelt, wenngleich sie in der vorliegenden Arbeit bei 20 von 50 Pferden kulturell nachgewiesen werden konnten. Eine Behandlung mit antibakteriell wirksamen Medikamenten allein gegen aerobe und fakultativ anaerob wachsende Bakterien, wie sie vorberichtlich in 42 Fällen erfolgte, verschafft den Anaerobiern möglicherweise einen zusätzlichen Wachstumsvorteil.

Darüber hinaus haben auch die Gabe von sekretolytisch wirksamen Medikamenten sowie stattgehabte Spülungen der NHH einen möglichen Einfluss sowohl auf die mikrobiologische als auch auf die zytologische Zusammensetzung der Sekrete. Es ist davon auszugehen, dass sich das zytologische Bild dieser Sekrete sowie die Morphologie einzelner Entzündungszellen mit fortschreitender Chronizität und in Abhängigkeit von der bakteriellen Besiedlung dynamisch verändern. Der Einfluss dieser Variablen konnte aufgrund ihrer mannigfaltigen Komplexität in der vorliegenden Arbeit nicht eingehender verfolgt werden.

4.1.1.2 Gesunde Pferde

Zum Zeitpunkt der Untersuchung und Probenentnahme bestanden bei keinem der Pferde Hinweise auf eine aktuelle oder abgeheilte Nasennebenhöhlenerkrankung. Da der Vorbericht dieser Pferde zum Teil aber nur wenige Wochen zurückzuverfolgen war, konnten stattgehabte Sinusitiden, die die Studienergebnisse möglicherweise verändern, nicht komplett ausgeschlossen werden. Eine sichere Abklärung hätte in Form einer CT Untersuchung des Pferdeschädels stattfinden können. Aufgrund des unverhältnismäßig großen Aufwandes wurde davon abgesehen.

4.1.2 Probengewinnung

4.1.2.1 Transendoskopische Probenentnahme

Der in dieser Arbeit für die transendoskopische Probenentnahme genutzte, über einen Handgriff und Führungsdraht steuerbare Katheter fand hier erstmals Anwendung in der Pferdemedizin. Der natürliche, transnasale Zugang birgt das Risiko der Kontamination aus dem Nüstern- und Nasengangbereich der gesammelten Proben, insbesondere vor dem Hintergrund, dass diese im Weiteren einer mikrobiologischen Untersuchung unterzogen werden. Um diesen Einflussfaktor zu minimieren, wurde der Katheter erst am Entnahmeort im Drainagewinkel der *Apertura nasomaxillaris* vorgeschoben und die Probe entnommen, bevor die endoskopische Untersuchung der Atemwege fortgeführt wurde. Nicht in allen Fällen konnte sichergestellt werden, dass bei der Probenentnahme allein aus den NNH abfließendes Sekret gewonnen wurde, vor allem wenn die Sekrete von sehr geringer Menge waren oder der Bereich um die Apertura nasomaxillaris stark zugeschwollen oder ganz verlegt war. Bei diesen Pferden wurde der Katheter, sofern aufgrund der veränderten anatomischen Verhältnisse möglich, in dem Bereich zwischen dorsaler und ventraler Nasenmuschel platziert, wodurch eine Kontamination mit Sekreten und respiratorischem Epithel der Nasenschleimhaut nicht vermieden werden konnte. Sekrete hoher Viskosität, wie es bei chronischen Sinusitiden der Fall war, die mit stark eingedicktem Eiter einhergingen, ließen sich mitunter nur schlecht in den kleinlumigen Katheter aspirieren.

Dennoch ließ sich mithilfe des flexiblen Katheters beim Großteil der Probanden zielgerichtet und nicht invasiv transendoskopisch eine Sekretprobe mit überschaubarem Kontaminationsrisiko entnehmen.

4.1.2.2 Probenentnahme nach chirurgischem Eingriff

Die Trepanation und Probenentnahme der Pferde der Kontrollgruppe erfolgte in Allgemeinanästhesie in Seitenlage und nach steriler Vorbereitung, wodurch das Kontaminationsrisiko so weit wie möglich reduziert wurde. Bei drei Pferden kam es jedoch zu geringgradigen Einblutungen in die Nasennebenhöhlen nach deren Eröffnung, wodurch die zytologischen Ausstriche mit stark erhöhter Anzahl von Erythrozyten auffielen. Eine Fehlinterpretation der zytologischen Ausstriche konnte allerdings vermeiden werden, da die Herkunft des Blutes bekannt war.

Die Trepanation und Probenentnahme der erkrankten Pferde erfolgte am stehenden, sedierten und lokalanästhesierten Patienten. Durch die aufrechte Haltung des Kopfes konnten hier Einblutungen von der Trepanationsöffnung in den Sinus vor der Probengewinnung besser verhindert werden und so eine Verfälschung der Untersuchungsergebnisse vermieden werden.

4.1.3 Probenauswertung

4.1.3.1 Auswertung der mikrobiologischen Proben

Die Auswertung aller mikrobiologischen Proben fand im Institut für Mikrobiologie der Stiftung Tierärztliche Hochschule Hannover gemäß den Vorgaben des Instituts für Probenaufbereitung und -untersuchung statt. Die Einteilung des Grades des bakteriellen Nachweises wurde semiquantitativ von verschiedenen Personen vorgenommen und unterlag daher zu gewissem Grad einer subjektiven Einschätzung. Eine computergestützte Auswertung würde hier eine noch objektivere Auswertung erlauben, wurde aber aufgrund des methodischen und finanziellen Mehraufwandes in dieser Studie nicht angewandt.

4.1.3.2 Auswertung der zytologischen Proben

Die Auswertung der zytologischen Proben erfolgte lichtmikroskopisch durch zwei unabhängige Personen in jeweils zwei zeitlich voneinander getrennten Durchläufen. Während die eine auswertende Person seit vielen Jahren Erfahrung mit der Bewertung von zytologischen Ausstrichen, insbesondere von Sekreten der tiefen Atemwege hat, hat sich die andere Person im Rahmen dieser Studie erstmals in dieses Gebiet eingearbeitet.

Die Interrater-Übereinstimmung war insgesamt lediglich zufriedenstellend, wobei höhere kappa-Werte und damit bessere Übereinstimmungen für Zellen erzielt wurden, die von beiden auswertenden Personen mit höheren Scorewerten bewertet wurden. Unabhängig davon wurden allerdings in den wiederholten Auswertungen häufig die gleichen signifikanten Unterschiede der zytologischen Befunde bei erkrankten Pferden und Proben der Kontrollgruppe herausgestellt. Daher wird die mikroskopische Auswertung von Sekreten der Nasennebenhöhlen als schnell zu erlernen und das Ergebnis der zytologischen Untersuchungen als gut reproduzierbar und verlässlich eingestuft. Die Intrarater-Übereinstimmung spiegelt mit deutlich höheren kappa-Werten diese Reproduzierbarkeit wider.

4.2 Diskussion der Ergebnisse

4.2.1 Ergebnisse der mikrobiologischen Untersuchung

Alle Proben, die transendoskopisch bei erkrankten Pferden entnommen wurden, wiesen bakterielles Keimwachstum auf, wohingegen in 24,7% (12 von 45) der transendoskopisch gewonnenen Kontrollproben von gesunden Pferden und der kontralateralen gesunden Kontrollseite erkrankter Pferde kein bakterieller Keimgehalt nachgewiesen werden konnte. Auch sechs von zehn nach Trepanation aus gesunden maxillären Sinus entnommene Proben blieben ohne Keimnachweis, während die übrigen vier grampositive Bakterien in geringgradiger Anzahl enthielten, denen aber keine spezielle pathogenetische Rolle bei Atemwegserkrankungen des Pferdes zugesprochen wird. Diese Ergebnisse sprechen entweder dafür, dass die equinen NNH im physiologischen Zustand kein oder nur ein kleines Spektrum an nachweisbaren Bakterien enthalten. Andererseits war bei allen Kontrollproben auch insgesamt nur sehr wenig Material für die Auswertung gewinnbar, sodass auch falsch-negative Ergebnisse aufgrund des geringen Probenvolumens und der damit verbundenen schlechteren Anzüchtbarkeit nicht auszuschließen sind.

4.2.1.1 Primäre Sinusitis

Streptococcus equi ssp. *zooepidemicus* wurde in 78,6% (11 von 14) der Pferde mit primärer Sinusitis nachgewiesen. Bei zehn dieser Proben war dieser Nachweis mittel- bis hochgradig, wovon es sich bei wiederum drei Proben um Reinkulturen han-

delte. Somit wurde dieser Keim bei primären Sinusitiden hochsignifikant häufiger isoliert als bei dentogenen Sinusitiden, bei denen er fast ausschließlich in Mischkulturen und vorrangig in geringgradigem Maße vorlag. Diese Ergebnisse zeigen eine gute Übereinstimmung mit Einzelnachweisen von *S. zooepidemicus* bei primären Sinusitiden (SCHUMACHER et al. 1987; TREMAINE u. DIXON 2001). Der Nachweis in drei Proben von klinisch und makroskopisch gesunden Kontrollseiten erkrankter Pferde lässt auf eine Kontamination von der an primärer Sinusitis erkrankten Seite schließen. Diese Kontamination kann unter Umständen durch die unterschiedlichen Druckverhältnisse im Nasopharynx während der In- und Exspiration zustande gekommen sein oder im Rahmen der Probengewinnung und -bearbeitung. Der Verdacht der Kontamination wird darin bestärkt, dass in keiner direkt aus einem gesunden Sinus entnommenen Probe *S. zooepidemicus* nachgewiesen werden konnte. Das Auftreten von *S. zooepidemicus* in mittelgradigem Keimgehalt bei zwei Fällen maligner Tumoren und einer Sinuszyste – einer Neoplasie, die im Inneren als primär bakteriologisch steril beschrieben wird (TREMAINE et al. 1999) – unterstreicht die Bedeutsamkeit radiologischer Verfahren, um raumfordernde Prozesse der NNH differentialdiagnostisch abzuklären. Zeitgleich wird darin die Limitation der mikrobiologischen Untersuchung als alleinstehendes Diagnostikum deutlich.
Weitere Erreger, die im Zusammenhang mit einer primären Sinusitis nachgewiesen werden konnten, entsprachen vorrangig bekannten sowohl grampositiven als auch gramnegativen Eitererregern des aeroben und fakultativ anaeroben Spektrums.

4.2.1.2 Dentogene Sinusitis

Sowohl in der Human- als auch in der Pferdemedizin ist die Beteiligung obligater Anaerobier an dentogenen Sinusitiden beschrieben worden (MACKINTOSH u. COLLES 1987; RUGGLES et al. 1993; BARTMANN et al. 2002; BROOK 2005; MEHRA u. JEONG 2009). Obligate Anaerobier konnten auch in dieser Studie bei der Mehrheit der dentogenen Sinusitiden nachgewiesen werden, wobei die Inzidenz bei der direkten Probenentnahme nach Trepanation höher war als bei der transendoskopischen Probenentnahme allein. Ihr semiquantitativer Nachweis war in der eigenen Studie stets hochgradig, und meistens lag mehr als ein anaerobes Isolat vor. Am weitaus häufigsten wurden *Fusobacterium* spp. (13 von 15 Proben) isoliert, die zuvor sowohl in der gesunden equinen Maulhöhle (BAILEY u. LOVE 1991), bei Zahnerkrankungen (BIENERT et al. 2003; KERN et al. 2016) als auch bei dentogenen Sinusitiden (BARTMANN et al. 2002) nachgewiesen worden sind. Weitere häufig isolierte obligate Anaerobier waren *Porphyromonas* spp., *Bacteroides* spp. und *Prevotella* spp..

Nur bei einem von vierzehn Pferden, bei denen eine primäre Sinusitis diagnostiziert wurde, wurden kulturell hochgradig Anaerobier nachgewiesen. Diese signifikant geringere Inzidenz von Anaerobiern bei primärer Sinusitis gegenüber dentogenen Sinusitiden spiegelt sich auch in der Literatur wider: Bisher wurde nur ein Fall einer primären Sinusitis mit Beteiligung anaerober Bakterien veröffentlicht (SCHUMACHER et al. 1987). Im Nachhinein ist die klinische Diagnose der primären Sinusitis mit hochgradigem Befall von Anaerobiern in dem einen Fall der vorliegenden Studie allerdings aus mehreren Gründen als fraglich anzusehen:

- Bei keinem der gesunden Pferde wurden Anaerobier in den Nasennebenhöhlen nachgewiesen, was eine endogene primäre Infektion der Schleimhaut unwahrscheinlich macht.
- Nachgewiesenermaßen können Anaerobier aber endogene Infektionen in der Maulhöhle verursachen (BIENERT et al. 2003), sodass ihr Nachweis in den Sinus wahrscheinlich an eine pathologische Verbindung zur Maulhöhle gekoppelt ist. Ein hämatogener Infektionsweg in die NNH gilt als wenig wahrscheinlich, auch wenn dieser für andere Organe wie Herz und Lunge beschrieben ist (VERDEGAAL et al. 2006) und eine Bakteriämie mit Anaerobiern nach Zahnextraktionen nachgewiesen wurde (KERN et al. 2016).
- Zu guter Letzt wurde in dem vorliegenden Fall keine computertomographische Untersuchung durchgeführt – eine sekundäre Sinusitis kann also trotz unauffälliger maulhöhlenendoskopischer und röntgenologischer Untersuchung nicht mit letzter Sicherheit ausgeschlossen werden.
- Nachfolgende Untersuchungen konnten aufgrund eines perakuten Kolikgeschehens, infolge dessen das Pferd euthanasiert wurde, nicht erfolgen.

Anaerobier traten zudem bei einem Pferd auf, bei dem histopathologisch ein Plattenepithelkarzinom in den Nasennebenhöhlen nachgewiesen wurde. Vorberichtlich wurden bei diesem Pferd drei Monate vor Klinikvorstellung auf der ipsilateralen Seite zwei Oberkieferbackenzähne extrahiert, sodass auch hier eine pathologische Verbindung zur Maulhöhle bestanden haben kann. Zwei weitere Fälle maligner Tumoren der NHH blieben ohne Nachweis anaerober Bakterien – bei diesen konnten im Rahmen der Untersuchungen auch keine orosinuidalen Fisteln festgestellt werden.

Die Ergebnisse der mikrobiologischen Untersuchungen haben gezeigt, dass sie bei primären und dentogenen Sinusitiden einen Hinweis auf die zugrundeliegende Erkrankung liefern können. Daher kann die transendoskopische Entnahme von Sekretproben aus dem Bereich der Apertura nasomaxillaris für mikrobiologische Untersuchungen als leicht durchführbares und hilfreiches diagnostisches Mittel angesehen werden. Nichtsdestotrotz bedarf es weiterer bildgebender Verfahren, um andere, in erster Linie neoplastische Prozesse auszuschließen. Zu Bedenken bleibt außerdem, dass sich durch Abwesenheit charakteristischer Sekretspuren im Drainagewinkel der NNH eine Füllung derselben nicht ausschließen lässt und unter diesen Umständen eine Probe nicht in jedem Fall sicher entnehmbar ist, sodass falschnegative Ergebnisse entstehen können.

4.2.1.3 Transendoskopische und direkte Probenentnahme im Vergleich

In drei Fällen von dentogener Sinusitis wurden Anaerobier erst in der direkten Probe aus den NNH nach Trepanation nachgewiesen. In diesen Fällen war die transendoskopische Probe nicht ausreichend, um die pathogenen Keime nachzuweisen. Eine Ursache hierfür kann der sehr heterogene Charakter der Nasennebenhöhlensekrete mit dünnflüssigen und festeren oder auch stark eingedickten Bestandteilen sein, die mit zunehmender Chronizität einen verkäsenden Charakter annehmen und nicht mehr über die natürlichen Drainagewege abfließen können. Ein Sistieren des Nasenausflusses kann in diesen beschriebenen Fällen zu einer Fehlinterpretation führen.

Auch eine vollständige Verlegung der sinunasalen Kommunikationswege bei hochgradiger Sinusitis kann mit fehlendem Abfluss gestauter Sekrete innerhalb der NHH einhergehen, die dann erst durch einen knöchernen Zugang erreicht werden können. Davon abgesehen bestand eine gute Verlässlichkeit darin, die pathogenen Erreger der Sinusitis in der transendoskopischen Probe nachzuweisen. Bei multiplem Keimnachweis von bis zu neun bakteriellen Isolaten innerhalb einer Probe konnte jedoch keine 100 prozentige Übereinstimmung der Probenpaare festgestellt werden, was sicherlich zum Teil auf den heterogenen Charakter der Nasennebenhöhlensekrete zurückzuführen ist.

4.2.2 Ergebnisse der zytologischen Untersuchung

Die Ergebnisse der zytologischen Auswertungen haben gezeigt, dass sich die gehäuft auftretenden Zellpopulationen für die verschiedenen untersuchten Gruppen der Pferde mit primärer und dentogener Sinusitis sowie gegenüber den Proben von gesunden NNH unterscheiden. Der bestimmende Zelltyp, der die Abgrenzung einer Probe eines erkrankten Pferdes von einer Kontrollprobe ermöglichte, waren neutrophile Granulozyten. Die Kontrollproben wurden von dem Auftreten von Epithelzellen bestimmt. Hier stand bei der Gewinnung meist nur eine sehr geringgradige Menge von serösen Sekreten zur Verfügung, sodass das Risiko der Verletzung des Epithels und somit dessen Abtragung größer war als bei den Proben erkrankter NNH mit viel Sekret. Diese Feststellung betont, dass die zytologische Auswertung die Diagnostik bei Sinusitis ergänzen kann, aber andere etablierte diagnostische Verfahren keinesfalls ersetzt.

Während in dieser Studie die zytologischen Charakteristika von Nasennebenhöhlensekreten beim Pferd erstmalig beschrieben worden sind, zählt bei Erkrankungen des unteren Respirationstrakts die zytologische Analyse von BALF oder TBS schon seit einigen Jahrzehnten zu den etablierten diagnostischen Verfahren (HOFFMAN 2008). Die Zytologie dient zum einen der Eingrenzung einer klinischen Verdachtsdiagnose. Des Weiteren ist beim equinen Asthma anhand des Gehalts an neutrophilen Granulozyten in der BALF ferner eine Abgrenzung einer entzündlichen Atemwegserkrankung (inflammatory airway disease, IAD) gegenüber der rekurrierenden Atemwegsobstruktion (recurrent airway obstruction, RAO) möglich (COUËTIL et al. 2016). Eine sichere ätiologische Einordnung der Sinusitiden in der vorliegenden Studie konnte mithilfe der Bestimmung des semiquantitativen Zellgehalts allerdings nicht erfolgen. Jedoch unterschieden sich das Auftreten und die Fülle der Bakterien bei dentogenen Sinusitiden deutlich von den übrigen untersuchten Ätiologien, sodass die Quantifizierung dieses Parameters mit wenig Zeitaufwand den Verdacht auf eine dentogen bedingte Sinusitis unter Umständen bestätigen kann.

4.2.3 Diagnostischer Nutzen der Sekretuntersuchung

Zusammenfassend haben die Ergebnisse dieser Studie gezeigt, dass sowohl die mikrobiologische als auch die zytologische Untersuchung von Sekreten aus den Nasennebenhöhlen des Pferdes als zusätzliches Diagnostikum dienen können, um erkrankte von gesunden NNH zu unterscheiden und um die ätiologische Diagnose einer Sinusitis bis zu einem gewissen Grad abzusichern.

Denkbar ist der Nutzen insbesondere bei Patienten, auf die folgende Befunde zutreffen:

- Einseitiger, purulenter Nasenausfluss, der nicht oder nur dezent übelriechend ist und endoskopisch sicher als von den NHH ausgehend festgestellt werden kann
- Im Rahmen einer sorgfältigen Maulhöhlenuntersuchung keine feststellbaren Pathologien der Zähne, die auf eine Infektion des Zahnwurzelbereichs, der Alveole oder des Zahnhalteapparates schließen lassen
- Röntgenologisch keine oder nicht eindeutige Befunde an den Backenzähnen, deren Wurzelbereiche in unmittelbarer Nähe zu den NNH liegen

Sollte bei Patienten mit diesen Befunden eine dentogene Ursache für die Sinusitis nicht sicher ausgeschlossen werden können, wäre die CT ein sinnvolles zusätzliches Diagnostikum. Durch die Sekretuntersuchung kann in einem solchen Fall möglicherweise die eindeutige Indikation für eine CT Untersuchung gestellt (bei Hinweisen auf eine dentogene Sinusitis) oder zunächst abgelehnt werden (bei Hinweisen auf eine primäre Sinusitis). Da der Kostenaufwand und bei einer CT in Allgemeinanästhesie das allgemeine Risiko für das Tier durchaus höher sind, könnte in diesen oben beschriebenen speziellen Fällen durch die Sekretuntersuchung eine CT Untersuchung möglicherweise vermieden werden.

Für andere sekundäre Sinusitiden, deren Ursprung nicht dentogen ist, waren die vorliegenden Fallzahlen für eine statistische Auswertung zu gering.

Bei einem der zwei Pferde mit einem PSH konnten zytologisch intrazytoplasmatische Pigmente festgestellt werden. Möglicherweise lassen sich diese in einer Spezialfärbung (Berliner-Blau-Färbung) als phagozytiertes Hämosiderin nachweisen, wie es bei Pferden mit belastungsinduziertem Lungenbluten der Fall sein kann (DOUCET u. VIEL 2002). Bei Fehlen von intrazytoplasmatischem Hämosiderin bei traumatisch bedingten Sinusitiden wäre so eine Differenzierung zu einem PSH anhand der zytologischen Untersuchung denkbar. Insbesondere bei malignen Neoplasien könnten gegebenenfalls im Sekretausstrich Tumorzellen identifiziert werden. Die Kasuistik der vorliegenden Studie erlaubt dazu allerdings noch keine ausreichende Aussage. Anders als bei primären oder dentogenen Sinusitiden, ist die Zusammenfassung aus Vorbericht, Charakter des Nasenausflusses, Befunde der speziellen Untersuchung des Kopfes, der Endoskopie und des Röntgens in den meisten Fällen bereits hinweisend auf die zugrundeliegende Erkrankung bei Vorliegen einer Gewebsneubildung (maligne und benigne Neoplasien, PSH, Sinuszysten) oder traumatischen Ursache.

Seitdem mithilfe genetischer Sequenzierung die Bestimmung des Mikrobioms von Organsystemen und Körperregionen möglich geworden ist, findet diese Methode vor allem in der Human- aber auch in der Tiermedizin Anwendung. Durch eine größere Genauigkeit und Sensitivität sind die kultur-unabhängigen Verfahren den konventionellen mikrobiologischen Anzuchtverfahren überlegen (FEAZEL et al. 2012; WILSON u. HAMILOS 2014). Auch wenn die Bestimmung des nasalen Mikrobioms bei gesunden und erkrankten Hunden und Katzen ebenso wie die des Mikrobioms der Maulhöhle des Pferdes in der Tiermedizin bereits durchgeführt wurde, so gibt es bislang noch keine Ansätze, diese Verfahren auch für die Bestimmung des Mikrobioms der Nasen- und Nasennebenhöhlen des Pferdes zu nutzen. Die vorliegende Studie weiterführend könnte jedoch mithilfe der Sequenzierung der bakteriellen 16S rDNA wahrscheinlich auch das Mikrobiom der equinen Nasennebenhöhlen in gesundem und erkranktem Zustand bestimmt werden. Gegebenenfalls ließen sich dadurch Veränderungen in der bakteriellen Besiedlung und Wechselwirkungen der Bakterien im Entstehungsprozess einer Sinusitis genauer charakterisieren.

5 Zusammenfassung

Hauke Gergeleit, Hannover (2019)

Zytologische und Bakteriologische Untersuchungen von Sekreten der Sinus paranasales des Pferdes

Einseitiger Nasenausfluss ist das häufigste Symptom bei Erkrankungen der Nasennebenhöhlen des Pferdes. Eine ätiologische Diagnose über die zugrundeliegende Erkrankung kann in vielen Fällen erst im Rahmen einer Kombination aufwendiger bildgebender Verfahren gestellt werden. Das primäre Ziel dieser Studie war die Beschreibung mikrobiologischer und zytologischer Befunde von Sekreten der Nasennebenhöhlen bei gesunden Pferden und Pferden mit unterschiedlichen Formen einer Sinusitis. Im Weiteren sollte der potentielle diagnostische Nutzen dieser Untersuchungsmethoden geprüft werden.

Im Rahmen von klinischen, röntgenologischen, endoskopischen und computertomographischen Untersuchungen sowie Maulhöhlenuntersuchungen wurden bei 50 Pferden mit einseitigem Nasenausfluss unterschiedliche Ätiologien von Sinusitiden diagnostiziert. Bei diesen Pferden wurden transendoskopisch mithilfe eines flexiblen, manuell steuerbaren Katheters Sekretproben aus dem Drainagewinkel der *Apertura nasomaxillaris* entnommen. Bei neunzehn Pferden konnte nach chirurgischer Eröffnung zu therapeutischen Zwecken zusätzlich eine Probe direkt aus den Nasennebenhöhlen entnommen werden. Zehn gesunde Kontrollpferde wurden auf die gleiche Weise transendoskopisch und nach Trepanation beprobt. Außerdem wurden bei 35 Pferden der Probandengruppe transendoskopisch Kontrollproben der kontralateralen, klinisch gesunden Kopfseite gewonnen. Alle Proben wurden mikrobiologisch und zytologisch untersucht.

Die mikrobiologische Untersuchung der Proben ergab ein hochsignifikant höheres Auftreten von *Streptococcus equi* ssp. *zooepidemicus* bei Pferden mit primärer Sinusitis, als bei Pferden mit dentogener Sinusitis oder gesunden Kontrollpferden, wohingegen Anaerobier hochsignifikant häufiger mit dentogenen Sinusitiden assoziiert waren. Von allen untersuchten Pferden wiesen solche mit dentogener Sinusitis zudem die höchste Diversität an Keimen auf.

Die zytologische Auswertung der Sekretproben wurde lichtmikroskopisch an nach Pappenheim gefärbten Ausstrichen von zwei Untersuchern vorgenommen. Neutrophile Granulozyten waren bei allen Formen der Sinusitis der dominierende Zelltyp und hatten häufig degenerativen Charakter. Zellen in Lyse traten insbesondere dann auf, wenn auch Bakterien lichtmikroskopisch nachgewiesen wurden. Zudem wurden häufig Epithelzellen und seltener Makrophagen und andere Entzündungszellen gefunden. Das Auftreten von extra- und intrazellulären Bakterien war häufiger und in größerem Ausmaß bei dentogenen als bei anderen Sinusitiden zu beobachten.

Die Ergebnisse haben gezeigt, dass die mikrobiologische Untersuchung von transendoskopisch gewonnenen Sekretproben der equinen Nasennebenhöhlen eine Abgrenzung von dentogenen Sinusitiden gegenüber anderen Formen der Sinusitis, insbesondere der primären Sinusitis, ermöglichen kann. Die zytologische Untersuchung dieser Sekrete ermöglicht die Differenzierung von gesunden und erkrankten Sinus paranasales. In bestimmten Fällen ist ferner eine Abgrenzung dentogener gegenüber anderen Sinusitiden möglich. Fallzahlbedingt konnte zur Diagnostik von anderen Erkrankungen der NNH über die Sekretanalyse in der vorliegenden Arbeit keine abschließende Aussage getroffen werden.

Die mikrobiologische und zytologische Sekretuntersuchung ist geeignet, die etablierten bildgebenden Verfahren zu ergänzen, um eine ätiologische Verdachtsdiagnose abzusichern, kann diese aber nicht ersetzen.

6 Summary

Hauke Gergeleit, Hannover (2019)

Cytological and microbiological examination of secretions from the paranasal sinuses in horses

Unilateral nasal discharge is the most common clinical symptom of paranasal sinus diseases in horses. An exact etiological diagnosis of the underlying cause can often only be achieved by a combination of several diagnostic imaging tools only.
The primary aim of this study was to describe the cytological and bacteriological features of secretions from the paranasal sinuses in healthy horses and horses with different forms of sinusitis. Furthermore, the ancillary diagnostic value of the examination of these secretions was assessed.

Fifty horses with a history of unilateral nasal discharge were included. Following clinical, radiographic, endoscopic and computed tomography examinations as well as examinations of the oral cavity, the horses were diagnosed with a specific paranasal sinus disease. In these horses the secretion samples were transendoscopically collected from the drainage angle of the nasomaxillary aperture by using a wire- guided, flexible catheter. In nineteen horses that underwent sinus surgery for therapeutic reasons, samples could be taken directly from the affected sinus. Samples were also collected from ten healthy control horses during endoscopic examination and following trephination. Additionally, in thirty-five diseased horses samples were taken transendoscopically from the contralateral side that was clinically unaffected. All samples were examined microbiologically and cytologically.

Streptococcus equi ssp. *zooepidemicus* was isolated significantly more in primary sinusitis than in dental sinusitis and healthy sinuses, whereas anaerobic bacteria were associated significantly more often with dental sinusitis. Furthermore, sinusitis with a dental origin showed a significantly higher diversity in the microbiological profile than all other horses.

For the cytological examination, the secretion samples were stained using the '*Pappenheim*' stain and evaluated by light-microscopy by two observers. Neutrophilic granulocytes were the dominating cell type in all forms of sinusitis and these cells often showed a degenerative character. Cells in lysis were closely associated with the detection of bacteria by light microscopy. Additionally, epithelial cells were detected frequently whereas macrophages and other inflammatory cells were found less often. Intra- and extracellular bacteria were found more often and in higher numbers in dental sinusitis than in other forms of sinusitis.

The results indicate that the microbiological examination of secretions from the paranasal sinuses that are taken under endoscopic guidance can help to distinguish between a dental sinusitis and other forms of sinusitis, especially primary sinusitis. The cytological examination of these secretions allows to differentiate between healthy and diseased paranasal sinuses. In specific cases a distinction of dental sinusitis from other forms of sinusitis is possible. Due to the small number of cases no final conclusion could be made regarding the diagnostic value of the analysis of secretions for other paranasal sinus diseases.

The microbiological and cytological examination of secretions provides an ancillary diagnostic tool in equine sinusitis, however, it cannot replace the established diagnostic imaging modalities.

7 Literaturverzeichnis

BAILEY, G. u. D. N. LOVE (1991):
Oral associated bacterial infection in horses: studies on the normal anaerobic flora from the pharyngeal tonsillar surface and its association with lower respiratory tract and paraoral infections.
Vet Microbiol 26, 367–379

BARAKZAI, S., H. TREMAINE u. P. DIXON (2006):
Use of scintigraphy for diagnosis of equine paranasal sinus disorders.
Vet Surg 35, 94–101

BARTMANN, C., M. PETERS, G. AMTSBERG u. E. DEEGEN (2002):
Dentogene Sinusitis durch gramnegative Anaerobier beim Pferd.
Tierärztl Prax Großtiere 30, 178–183

BEARD, W. u. J. HARDY (2001):
Diagnosis of conditions of the paranasal sinuses in the horse.
Equine Vet Educ 13, 265–273

BIENERT, A., C. BARTMANN, J. VERSPOHL u. E. DEEGEN (2003):
Bakteriologische Befunde bei endodontischen und apikalen Erkrankungen der Backenzähne des Pferdes.
Dtsch Tierarztl Wochenschr 110, 358–361

BROOK, I. (2005):
Microbiology of acute and chronic maxillary sinusitis associated with an odontogenic origin.
Laryngoscope 115, 823–825

COUËTIL, L., J. CARDWELL, V. GERBER, J. P. LAVOIE, R. LÉGUILLETTE u. E. RICHARD (2016):
Inflammatory airway disease of horses—revised consensus statement.
J Vet Intern Med 30, 503–515

COWELL, R. L. u. R. D. TYLER (2001):
Diagnostic Cytology and Hematology of the Horse E-Book.
Elsevier Health Sciences.

DERKSEN, F., C. BROWN, I. SONEA, B. DARIEN u. N. ROBINSON (1989):
Comparison of transtracheal aspirate and bronchoalveolar lavage cytology in 50 horses with chronic lung disease.
Equine Vet J 21, 23–26

DIXON, P. M., T. D. PARKIN, N. COLLINS, C. HAWKES, N. TOWNSEND, W. H. TREMAINE, G. FISHER, R. EALEY u. S. Z. BARAKZAI (2012):
Equine paranasal sinus disease: a long-term study of 200 cases (1997–2009): ancillary diagnostic findings and involvement of the various sinus compartments.
Equine Vet J 44, 267–271

DOUCET, M. Y. u. L. VIEL (2002):
Alveolar macrophage graded hemosiderin score from bronchoalveolar lavage in horses with exercise-induced pulmonary hemorrhage and controls.
J Vet Intern Med 16, 281–286

FEAZEL, L. M., C. E. ROBERTSON, V. R. RAMAKRISHNAN u. D. N. FRANK (2012):
Microbiome complexity and Staphylococcus aureus in chronic rhinosinusitis.
Laryngoscope 122, 467–472

HENNINGER, W., E. MAIRI FRAME, M. WILLMANN, H. SIMHOFER, D. MALLECZEK, S. M. KNEISSL u. E. MAYRHOFER (2003):
CT features of alveolitis and sinusitis in horses.
Vet Radiol Ultrasound 44, 269–276

HOFFMAN, A. M. (2008):
Bronchoalveolar lavage: sampling technique and guidelines for cytologic preparation and interpretation.
Vet Clin North Am Equine Pract 24, 423–435

KAMINSKY, J., A. BIENERT-ZEIT, M. HELLIGE, K. ROHN u. B. OHNESORGE (2016):
Comparison of image quality and in vivo appearance of the normal equine nasal cavities and paranasal sinuses in computed tomography and high field (3.0 T) magnetic resonance imaging.
BMC Vet Res 12, 13

KERN, I., C. BARTMANN, J. VERSPOHL, J. ROHDE u. A. BIENERT-ZEIT (2016):
Bacteraemia before, during and after tooth extraction in horses in the absence of antimicrobial administration.
Equine Vet J 49, 178–182

MACKINTOSH, M. u. C. COLLES (1987):
Anaerobic bacteria associated with dental abscesses in the horse and donkey.
Equine Vet J 19, 360–362

MAY, A. u. H. GEHLEN (2009):
Durchführung, Analyse und Aussagekraft von Tracheobronchialsekret (TBS) und Bronchoalveolärer Lavage (BAL) bei Pferden mit Lungenerkrankungen.
Pferdeheilkunde 25, 310–320

MEHRA, P. u. D. JEONG (2009):
Maxillary sinusitis of odontogenic origin.
Curr Allergy Asthma Rep 9, 238–243

RUGGLES, A. J., M. W. ROSS u. D. E. FREEMAN (1993):
Endoscopic examination and treatment of paranasal sinus disease in 16 horses.
Vet Surg 22, 508–514

SCHUMACHER, J., C. HONNAS u. B. SMITH (1987):
Paranasal Sinusitis Complicated by Inspissated Exudate in the Ventral Conchal Sinus.
Vet Surg 16, 373–377

TREMAINE, W., C. CLARKE u. P. DIXON (1999):
Histopathological findings in equine sinonasal disorders.
Equine Vet J 31, 296–303

TREMAINE, W. u. P. DIXON (2001):
A long-term study of 277 cases of equine sinonasal disease. Part 1: Details of horses, historical, clinical and ancillary diagnostic findings.
Equine Vet J 33, 274–282

TREMAINE, W. u. D. E. FREEMAN (2007):
Disorders of the paranasal sinuses.
Equine respiratory medicine and surgery. Saunders Elsevier, Philadelphia 403–404

VERDEGAAL, E., N. D. HEER, N. MEERTENS, J. MAREE u. M. S. VAN OLDRUITENBORGH-OOSTERBAAN (2006):
A right-sided bacterial endocarditis of dental origin in a horse.
Equine Veterinary Education 18, 191–195

WAGUESPACK, R. W. u. J. TAINTOR (2011):
Paranasal sinus disease in horses.
Compend Contin Educ Vet 33, E1-E12

WILSON, M. T. u. D. L. HAMILOS (2014):
The nasal and sinus microbiome in health and disease.
Curr Allergy Asthma Rep 14, 485

8 Danksagung

Zunächst bedanke ich mich bei Herrn Prof. Dr. Bernhard Ohnesorge für die Möglichkeit, diese Doktorarbeit unter seiner Leitung durchführen zu können. Vielen Dank für die Bereitstellung des Themas, die wissenschaftliche Betreuung, die konstruktiven Gespräche sowie das Vertrauen, dass Sie mir stets entgegen gebracht haben.

Liebe Astrid, Du hast einen ganz wesentlichen Anteil zum Gelingen dieser Arbeit beigetragen. Ich habe mich von Dir zu jeder Zeit fantastisch betreut, beraten und unterstützt gefühlt. Binnen kürzester Zeit hast Du mit mir Lösungsansätze erarbeitet, Entwürfe korrigiert und immer ein offenes Ohr für meine Belange gehabt. Besser hätte ich es da vermutlich nicht treffen können. Dafür danke ich Dir sehr!

Liebe Franzi, Du warst von Beginn an meine seelische Unterstützung und mir immer eine sehr gute Freundin. Wir haben uns zusammen durch die vergangenen Jahre gekämpft, aber vor allem immer wieder gemeinsam gelacht, gesungen und getanzt. Ich bin sehr froh, diesen Abschnitt mit Dir bestritten zu haben und danke Dir von Herzen für viele schöne Momente und Deine Freundschaft!

Vielen Dank an das gesamte Team der Klinik für Pferde. Ohne die Unterstützung beim Probensammeln und Auswerten wäre diese Arbeit so nicht möglich gewesen. Danke für Euren Rückhalt über die tägliche Arbeit hinaus.

Das größte Dankeschön gilt meiner lieben Familie. Meinen Eltern, die mich jederzeit bedingungslos unterstützt und nie an mir gezweifelt haben. Mama, ich danke Dir für alles und widme Papa und Dir diese Arbeit!
Sabine und Stefan, danke dafür, dass Ihr die weltbesten Geschwister seid und wir dieses unschlagbare Team bilden! Danke an meine liebe Tante Marlies, die mir zeitlebens zur Seite steht. Danke, liebe Oma Herta, Dein Stolz und Deine Zuversicht beflügeln uns jeden Tag.

9 Erklärungen über die erbrachten Eigenleistungen

Gemäß § 8 Absatz 3 der Promotionsordnung der Tierärztlichen Hochschule Hannover hat der Promovent bei einer Dissertation, die auf Veröffentlichungen basiert (kumulative Dissertation), an denen mehrere Autoren beteiligt waren, den selbstständigen Anteil an den vorgelegten Publikationen darzulegen.

- Klinische Allgemeinuntersuchung, spezielle Untersuchung des Kopfes, Endoskopie der oberen Atemwege sowie Probenentnahmen bei allen an der Studie beteiligten Pferden
- Auswertung der Befunde weiterer bildgebender diagnostischer Mittel (Röntgen, CT)
- Zusammenstellen der Befunde aus den klinischen und bildgebenden Untersuchungen
- Zusammenstellen und Gruppieren der mikrobiologischen Befunde
- Erstellen und Aufbereitung der zytologischen Ausstriche
- Mikroskopische Auswertung der zytologischen Präparate als einer der zwei Untersucher
- Anfertigung lichtmikroskopischer Fotografien
- Auswahl und Aufbereitung repräsentativer lichtmikroskopischer Fotografien für die Veröffentlichung
- Anfertigen der Publikationen